Secondary Ion Mass Spectrometry and Its Application to Materials Science (Second Edition)

Online at: https://doi.org/10.1088/978-0-7503-3331-3

Secondary Ion Mass Spectrometry and Its Application to Materials Science (Second Edition)

Sarah Fearn
Department of Materials, Imperial College, London SW7 2AZ, UK

IOP Publishing, Bristol, UK

ISBN 978-0-7503-3331-3 (ebook)
ISBN 978-0-7503-3329-0 (print)
ISBN 978-0-7503-3332-0 (myPrint)
ISBN 978-0-7503-3330-6 (mobi)

DOI 10.1088/978-0-7503-3331-3

Version: 20250801

IOP ebooks

British Library Cataloguing-in-Publication Data: A catalogue record for this book is available from the British Library.

Published by IOP Publishing, wholly owned by The Institute of Physics, London

IOP Publishing, No.2 The Distillery, Glassfields, Avon Street, Bristol, BS2 0GR, UK

US Office: IOP Publishing, Inc., 190 North Independence Mall West, Suite 601, Philadelphia, PA 19106, USA

Contents

Preface

The main aim of this book is to provide an introduction and overview to secondary ion mass spectrometry (SIMS) instrumentation and its application as an analytical technique in material science characterisation. It has been written with the aim to inform and introduce the science behind the technique to final year undergraduates and researchers who may be starting to use secondary ion mass spectrometry in their research. The book provides an outline of the different ion beams used in the technique along with other important components of the instrument such as the mass analysers, and ion detectors. A clear description of the types of analyses that can be carried out and the data obtained are also given, along with a description of the beam interactions that occur during an analysis and need to be understood to accurately interpret the data. Finally some examples of the techniques applications to material characterisation are provided; as well as conventional analyses, different approaches are also presented to highlight the flexibility of the SIMS technique.

Author biography

Sarah Fearn

Sarah Fearn obtained her PhD in Materials Science and Engineering from the Department of Materials, Imperial College, University of London, in 2000. Her focus is the secondary ion mass spectrometry (SIMS) analysis of ultra-shallow implants in silicon. After working in commercial SIMS analysis for two years, she returned to academia in 2002, studying glass corrosion using SIMS. She has worked on a variety of research projects over the years that have all applied SIMS as an important characterisation tool for understanding a range of materials science issues. Over the last 20 years, she has published over 100 papers. Since 2012, she has been the manager of the Surface Analysis Facility in the Department of Materials at Imperial College.

Sarah Fearn

Chapter 1

Introduction

Secondary ion mass spectrometry (SIMS) has been at the forefront of high-resolution materials analysis and characterisation since the 1960s. A combination of factors makes SIMS unique among the analytical techniques widely available. It can detect all elements of the periodic table from H to U, along with isotopes and molecular species. Under optimum conditions, the technique has extremely high sensitivity down to ppm (and in some cases ppb) coupled with very high surface specificity on the order of nanometres and high lateral resolving powers down to 100 nm for some instrumentation. This range of capabilities means that SIMS has been exploited in many areas of research and materials development.

SIMS instrumentation has dramatically changed from the earliest spectrometers. These early instruments were once used for specific purposes: some were dedicated to depth profiling of materials using high ion beam currents to analyse the near-surface to bulk regions of materials (dynamic SIMS), while others, such as time-of-flight instruments, produced complex mass spectra of the very outermost surfaces of samples, using very low beam currents (static SIMS). Now, with the development of dual-beam instruments, the distinction between these two applications often disappears.

Fundamental changes have also occurred in the areas of materials where SIMS can now be usefully applied. For a very long time, subsurface information about organic materials could not be obtained via SIMS due to the fragmentation of the molecular structures caused by the ion beams. The development of cluster ion beam technology has now consigned this limitation to history, and depth profiles (as well as 3D analyses) can now be readily achieved. Similarly, early applications of SIMS in the biosciences highlighted its huge potential; as such, its application within this field of study has been a major factor in developing the technique further: adaptions of mass analysers and cryo-sample stages coupled with the introduction of advanced

doi:10.1088/978-0-7503-3331-3ch1

data analysis techniques to sift through the vast arrays of data produced have made SIMS a valuable tool in the biomedical arena.

There are now very few limitations on the samples that can be analysed by SIMS. The main caveat is that the sample must be vacuum compatible, but for the most part, this can be achieved. The wide variety of SIMS platforms and ion beam capabilities that are now available offer endless opportunities for the application of SIMS. All it requires is a little creativity, imagination, and patience.

IOP Publishing

Sarah Fearn

Chapter 2

Secondary ion generation, analysis, and detection

Mass spectrometry is one of the major characterisation techniques used in analytical science. It is based on the measurement of the mass-to-charge (*m*/*z*) ratio of ions. Any atomic or molecular species that can be ionised and transported into a gas phase can, in principle, be analysed by mass spectrometry, which therefore makes it a universal technique. A wide variety of different types of mass spectrometry exist, such as gas chromatography (GC–MS), high-performance liquid chromatography (HPLC-MS), inductively coupled plasma (ICP-MS), laser ablation inductively coupled plasma (LA-ICP-MS), matrix-assisted laser desorption/ionisation (MALDI), desorption electrospray ionisation (DESI-MS); secondary ion mass spectrometry (SIMS) is part of this family of techniques.

Practically, to carry out a SIMS analysis, three steps are generally required:

(1) secondary ion generation (ionisation) from the target material;
(2) analysis (mass separation) of the sputtered secondary ions from the target; and
(3) detection of the mass-separated secondary ions.

In the sections below, each of these stages will be described.

2.1 Secondary ion generation

Secondary ions can be generated via four physical methods: electron, chemical, spray, or desorption ionisation. In the first three methods, the original sample must be turned into either the gas phase (vaporised) or the liquid phase before the secondary ions are generated. In SIMS, the secondary ions are formed through the desorption ionisation of a sample using an energetic (primary) ion beam that is rastered over the sample surface. As the sample remains in its original solid state,

doi:10.1088/978-0-7503-3331-3ch2

chemico-spatial information can be obtained during a SIMS analysis. This is one of the major advantages SIMS has over many other analytical techniques where a change of sample state is required.

Primary ion beams that bombard a sample's surface can be produced in a number of ways and can be formed from either gases (e.g. oxygen or argon) or metals (e.g. caesium or bismuth) as the ion beam source material. Oxygen and caesium ion beams have been extensively used as ion sources for depth profiling materials due to their chemical effects that cause an enhancement of secondary ion signals. This is discussed in more detail in section 3.1.

Primary ion sources such as liquid metal ion guns or sources (LMIG/S) based on gallium or bismuth have become increasingly popular because the near-point ion source can be refocused to form a very small spot on the sample's surface. This has been successfully exploited in ion mapping and microprobe analysis, as the lateral resolving power of these ion sources can give submicrometre resolution. In the last decade, significant developments in ion beam technology, most notably gas cluster ion beams (GCIBs), have meant that SIMS can now be applied to the field of 'soft' organic materials. The advantage of the GCIB sources is their very low ion impact energy per ion, resulting in much less destructive analysis and less fragmentation of molecular signals. In the following section, an outline of the most prevalent ion beam sources currently in use today is presented.

2.1.1 Gas sources

An energetic primary ion beam can be formed from a gaseous element or molecule in a variety of ways. Common to these gas sources is the initial formation of a plasma. The charged gas ions are then drawn from the plasma and focussed down the ion beam column. The differences between gas sources lie in the way the initial plasma is formed. These differences result in primary ion beams with varying characteristics with regard to their energies, currents, focus, and brightness. The final application of the primary ion source, e.g. depth profiling or ion mapping, therefore dictates the most appropriate ion beam source to use.

2.1.1.1 Electron impact

The simplest of primary ion gas sources used in SIMS instrumentation is the electron impact (EI) source. Commonly, a tungsten filament set to an optimised current is positioned adjacent to a small ionisation chamber (see the schematic in figure 2.1). A selected gas such as O_2, Ar, or Xe is leaked into the chamber in a controlled manner. High-energy electrons emitted from the filament are drawn into the ionisation chamber and displace electrons upon impact with the gas atoms, resulting in the formation of positive gas ions. The positive gas ions are then extracted from the ionisation chamber for focussing.

The ion beams produced by these sources have large currents and can be used at energies that typically range from 250 eV to 10 keV. These sources tend to have poor beam focus, with full width at half maxima varying from several to tens of microns in size. These types of ion beams, therefore, are not ideal for SIMS experiments that

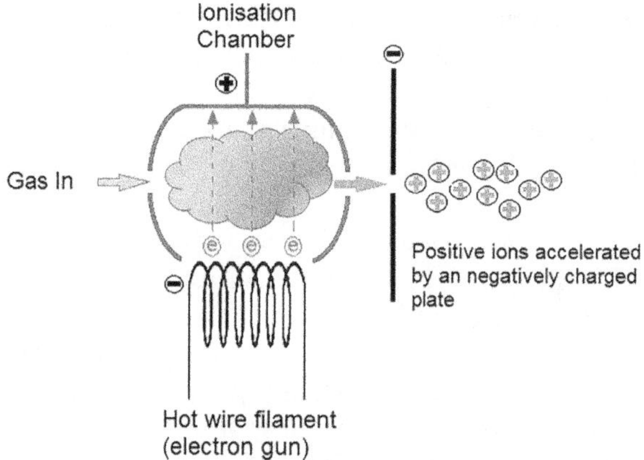

Figure 2.1. Schematic of an EI ionisation source.

require high lateral resolving power, e.g. ion mapping, but are efficient for the depth profiling of inorganic materials.

2.1.1.2 Duoplasmatron

A duoplasmatron ion source incorporates a cathode (typically nickel), an intermediate electrode, and an anode in its construction. The duoplasmatron can form a plasma through two modes of operation: the hot cathode mode and the cold cathode mode. In the hot cathode mode, a hot filament injects electrons into a gas, ionising it and forming a plasma. The hot cathode mode is used to form ion beams using inert gases such as argon and xenon. In the cold cathode mode of operation, an electric current is passed directly through the gas, creating a glow discharge. The resulting plasma is confined to a small space near the anode by a magnetic field. The magnetic field also helps to reduce the ion beam diameter and therefore has a net focusing effect. Primary ion beams of reactive gases such as O_2 and N_2 are formed using the cold cathode setup. A schematic of a duoplasmatron ion source is shown in figure 2.2.

Duoplasmatrons are highly efficient at ionising the gas species in the source. The ion beam energy spread is between 5 and 15 eV, and Ar^+ and O_2^+ ion beams typically have brightnesses ranging from 250 to 500 A m^{-2} Sr^{-1} V^{-1} [2].

2.1.1.3 RF plasma

Common direct current (DC) plasma ion sources have limited lifetimes due to the rate of cathode erosion that occurs during their operation. The lifetimes of a DC plasma ion source scale inversely with plasma density and source brightness. Ideally, a plasma source should have a long, predictable operational lifetime. Other important and desirable characteristics include high brightness, low energy spread (required for good focus), high beam purity, good current stability, and finally the option to select different ion beam species, i.e. inert gases such as neon or argon, as well as the more reactive oxygen gas source or a molecular gas.

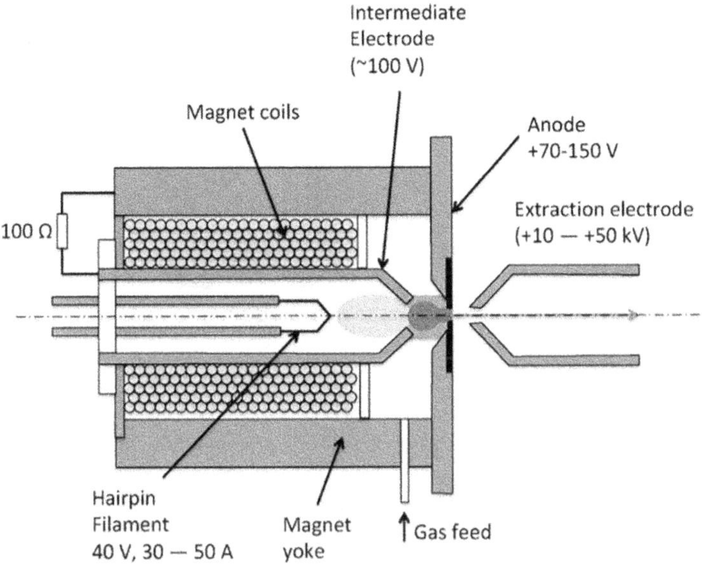

Figure 2.2. Schematic of a duoplasmatron ion source. Reproduced from [1].

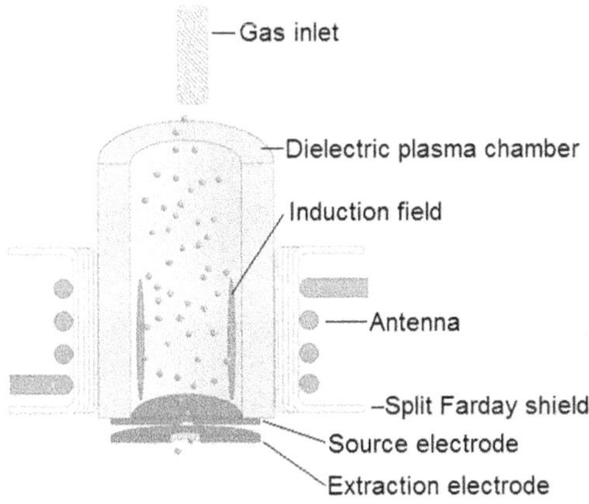

Figure 2.3. Schematic of an ICP ion beam source.

The development of inductively coupled plasma (ICP) ion sources (see figure 2.3) has provided many of these capabilities, along with the benefit of ion beams with fast milling rates and high lateral resolving power for imaging.

Along with improved ion beam lifetimes, current stabilities, and the ability to select other gases, ICP ion sources also have advantages associated with parameters such as brightness and focus, which expand their applications, as very high material mill rates can be achieved. For example, a typical 30 keV Ga^+ ion beam with a current of 20 nA can mill silicon at a rate of $\sim$5 $\mu m^3\ s^{-1}$ compared to a 30 keV Xe^+ ICP ion beam, which has mill rates of $\sim$100 $\mu m^3\ s^{-1}$ [3]. These high mill rates are useful for certain features of interest, such as interfaces buried at large depths in a sample or for the production of geological cross sections. In the case of oxygen-based ICP ion sources, the beam is highly focussed, with beam spot sizes of 50 nm compared to 200 nm in duoplasmatron ion sources.

2.1.1.4 Cluster ion beams

Monoatomic primary ion beam sources have been the dominant ion beam sources used in SIMS instruments for many decades. As previously mentioned, beneficial chemical effects can be achieved with oxygen and caesium ion beams, but the physical effects caused by the energetic ion impact on the target material have limited SIMS to mainly inorganic materials (see section 3.2 for more details on ion beam interactions). Even with the development of ion beams that operate at very low energies [4], impact energies are still high compared to the typical binding energy of ~10 eV for polymer-based materials. The resulting fragmentation of the molecular structure under ion beam bombardment means that stable signals related to the molecular structure of the material being analysed are not obtained. To this end, the development of cluster ion beams has largely overcome this issue.

Early experiments to develop cluster beams for SIMS were carried out with giant glycerol [5], SF_5^+ [6], and C_{60}^+ ion beams, which initially showed great promise for the depth profiling of polymeric materials [7]. In all cases, these early cluster ion beam sources showed that the impact energy per atom on the target material was dramatically reduced compared to that of monoatomic ion beams, causing less fragmentation of the molecules that are the foundation of the material (see section 3.2.4 for more details of ion beam cluster interactions). However, technical difficulties with these early sources meant that uptake was poor. As technical challenges were progressively overcome, Ar-based GCIBs have become readily available and currently appear to be the dominant cluster ion beam technology for molecular depth profiling [8].

Initially, novel nitrogen GCIBs were developed for GaN growth in the early 2000s [9], and adaptations of these early ion beam systems have led to the gas cluster ion sources commonly found in labs today. GCIBs are formed by the expansion of a gas through very narrow supersonic nozzles. The basic GCIB configuration is shown in the schematic of figure 2.4. The source requires a high level of vacuum pumping, and there are typically four chambers: a nozzle chamber, a differential pumping chamber, an ionisation/acceleration chamber, and a target chamber, with each chamber being individually pumped [10].

As mentioned earlier, a neutral gas cluster beam is generated in the nozzle chamber by passing a highly compressed gas at several atmospheres through the very small expansion nozzle. The clusters form due to the adiabatic expansion of the gas that occurs at the nozzle. After passing the skimmer, the neutral clusters are then positively ionised by electron bombardment and accelerated down the column by a series of electrodes. Cluster and monomer ions are separated and filtered by a strong magnetic field, and the cluster ion beam is focussed down the column by electrostatic lenses until the beam reaches the target.

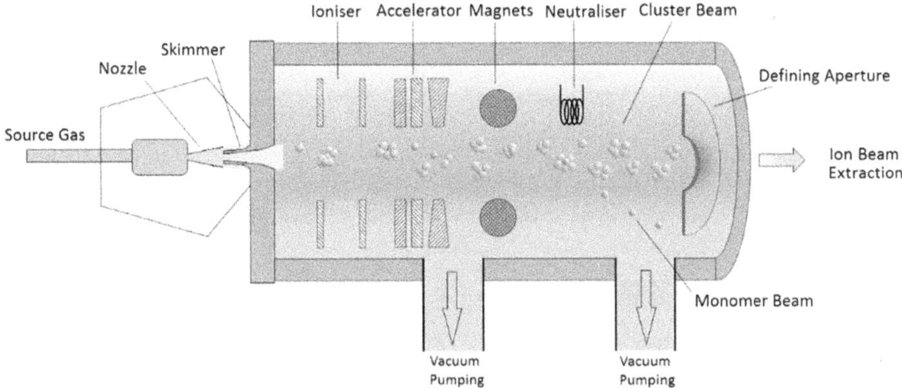

Figure 2.4. Schematic of a basic GCIB source.

2.1.2 Metal-based sources

Ion beam sources based on caesium are well established and widely used due to their favourable enhancements of negative secondary ion yields. Recently, liquid metal ion guns/sources (LMIG/Ss) similar to those used in focussed ion beam (FIB) systems have become common, primarily in time-of-flight-based SIMS instruments such as those produced by IONTOF. Some of the early LMIG/Ss used in SIMS instruments were based on gallium, then gold [11], but now bismuth ion sources are the predominant ion beam sources [12], as improved secondary ion yields are achieved with bismuth compared to gallium. This becomes even more noticeable when the small cluster form of Bi_3^+ is used [13]. Similarly to FIB systems, bismuth LMIG/Ss are extremely well focussed, which translates into high lateral resolving power, and they have been widely exploited in the area of ion beam mapping. As the beam is also typically operated at very low currents in time-of-flight-based SIMS instruments, ion mapping of biological materials has been very successful. In comparison, the more 'traditional' Cs-based ion sources are typically operated at much higher ion beam currents and are used primarily for the depth profiling mode of operation.

2.1.2.1 Cs sources

Based on caesium (Cs) or more typically a compound of caesium such as Cs_2CO_3, these sources form ion beams via thermal ionisation. A reservoir of Cs is electrically heated to a temperature sufficient to create a Cs vapour. The vapour strikes a metallic frit that is heated to ~1000 C. Upon striking the frit, the Cs vapour loses an electron (Cs has a very low work function) and becomes ionised. An extraction plate then draws the Cs^+ ions down into the primary column (see figure 2.5).

Caesium guns are efficient and can be used over a wide range of energies (from 500 eV to 10 keV) and ion beam currents. They are very widely used in the depth profiling of semiconductor materials, with a particular focus on phosphorus or arsenic implant profiles due to the enhanced negative secondary ion yields that are obtained for electronegative species. The ability to analyse at low beam energies also means that excellent depth resolutions can be obtained.

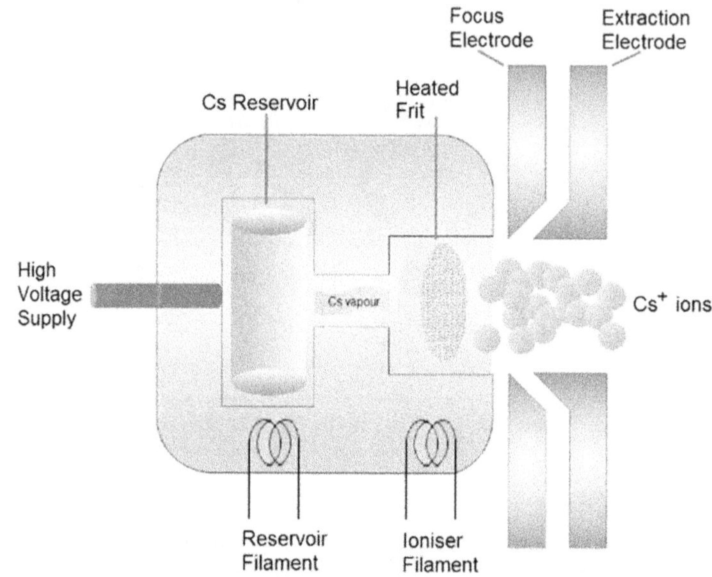

Figure 2.5. Schematic of a caesium surface ionisation source.

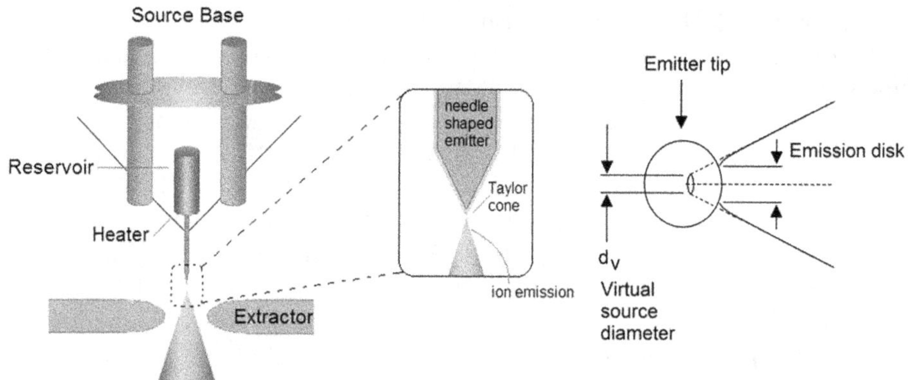

Figure 2.6. Schematic of a liquid metal ion gun. The diameter of the virtual source is d_v.

2.1.2.2 Liquid metal ion guns/sources

The LMIG/Ss used in SIMS instruments are based on the same technology as that found in gallium (Ga)-based FIB systems; however, for many time-of-flight SIMS (ToF-SIMS) instrument applications, bismuth (Bi) is typically the metal ion source material. A schematic of this type of source is shown in figure 2.6.

A Bi reservoir is slowly heated in a controlled manner by an electrical coil. Once the Bi becomes molten, it flows down a tungsten (W) needle to its tip, where opposing forces inside the molten metal, namely electrostatic, surface tension, and pressure forces, form a conical structure known as a Taylor cone at equilibrium. A very high electric field is applied to the end of the tip (10^8 V cm^{-1}), and at a critical voltage—the threshold voltage, V_s—ions are ejected from the tip of the Taylor cone.

The Bi atoms become ionised via field emission (with the resultant electrons flowing back to the ion source base to ground) and are then accelerated and focussed by lenses down the ion column to the target. As the Bi ions are extracted, the resulting emission current increases with increasing extraction voltage.

A figure of merit used to define an LMIG/S is brightness, β, defined by the equation:

$$\beta = \frac{4}{\pi} \frac{(\mathrm{d}I/\mathrm{d}\Omega)}{\mathrm{d}v^2} \left(\mathrm{A\ cm^{-2}\ sr^{-1}}\right), \tag{2.1}$$

where ($\mathrm{d}I/\mathrm{d}\Omega$) is the angular intensity of the ion beam measured in steradians, and d_v is the diameter of the virtual source. This is measured by extrapolating the tangents to the trajectories of the emitted ions, as highlighted in figure 2.6, and increases with the emitted ion current.

Liquid metal ion sources have very high brightness values, typically of the order of 10^6 A $\mathrm{cm^{-2}}$ $\mathrm{sr^{-1}}$, along with very low energy spreads (i.e. they are close to monochromatic). The high focus obtained using LMIG/Ss has enabled greater exploitation of the chemico-spatial power of SIMS, with the capability of resolving features below 1 μm in both inorganic and organic materials in ion imaging mode.

2.2 Filtering, focussing, and steering

Once the ion beam has been generated at the ion beam source, it still must be directed to and over the target material in order to create the secondary ions that are then measured.

2.2.1 Wien filter

The extracted primary ion beam in a SIMS instrument contains a few percent of impurity ions, i.e. ions of multiple charge states, including neutrals formed during or after the acceleration phase [14]. There are also multiply charged or polymer ions of the primary working gas. The ion beam needs, therefore, to be purified to ensure consistent bombardment conditions before striking the sample surface. This is carried out using a Wien filter, as shown in the schematic of figure 2.7. The Wien filter is placed in the ion beam column, whereby a perpendicular electric field, E, and a magnetic field, B, filter out the extraneous particles based on their velocities, resulting in a pure beam passing down the ion beam column. A slight bend in the ion beam column is used to separate out neutrals from the charged ions [15].

2.2.2 Beam focus

Primary ion beams are steered and focussed using electrostatic optics to produce a focal point on the sample. The focal spot typically has a much larger length scale than that of the primary ion impact zone. The shape and size of the focal spot depend upon the geometry of the instrument and the precise settings of all the lenses and steering plates. Unless particular care is taken, the focal spot is rarely a well-defined Gaussian shape and is often asymmetric. Nevertheless, because the beam is

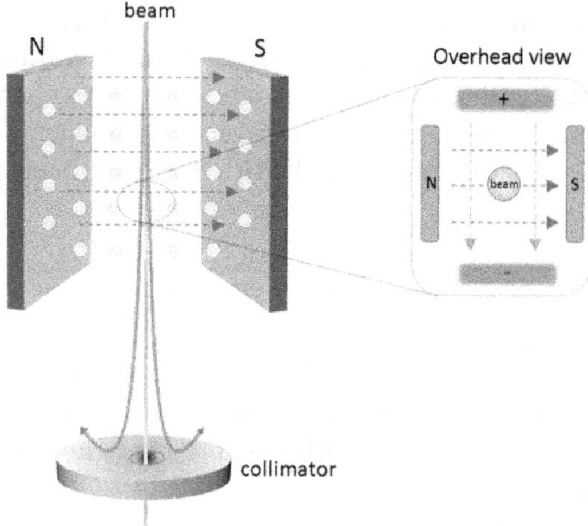

Figure 2.7. Schematic of a Wien filter. The primary input beam is purified by separating out the extraneous particles based on their velocities using perpendicular electric, *E*, and magnetic, *B*, fields highlighted in the overhead view.

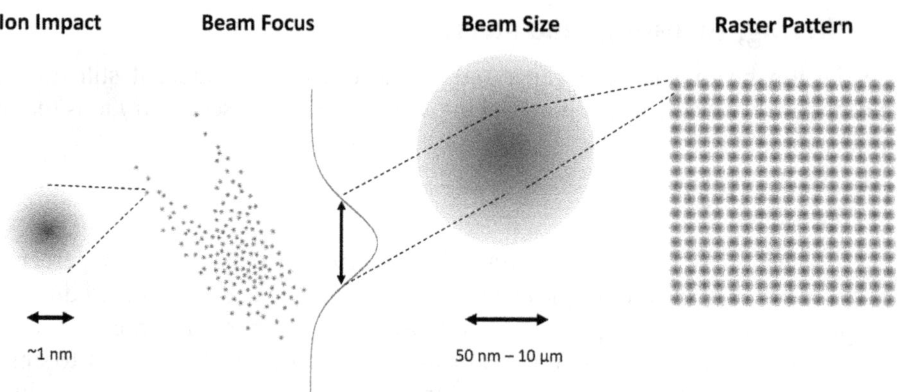

Figure 2.8. Comparison of lateral length scales and the relationships between individual primary ion impacts, beam focus, beam spot sizes, and beam raster patterns.

almost never stationary on the sample, it is convenient to consider it as having a symmetric shape with a characteristic beam diameter. The beam diameter is taken to be the full width at half maximum intensity of the average ion current distribution on the sample projected onto a line. The direction of the line is determined by the way in which the beam diameter is measured, which is often carried out by imaging a sharp straight edge on a reference sample. In some instruments designed for nanoscopic imaging, this characteristic diameter can be as small as a few tens of nanometres; however, it is typically $\sim$1 μm for standard analytical ion beams and more than 10 μm for sputtering ion beams. Figure 2.8 shows the relationship between ion impacts, beam focus, beam spot sizes, and beam raster patterns.

Secondary ion mass instruments that use LMIG/Ss as the primary ion beam source may also have different operational modes due to the way the primary source ions are drawn down the ion column. The schematic in figure 2.9 shows the three common modes available on some ToF-SIMS instruments: the high-current bunched mode (HCBM), the burst alignment mode (BAM), and the collimated mode. These different operational modes of the LMIG/S give rise to different analytical capabilities.

In the HCBM, a very short beam pulse (beam width) of primary ions, ~10 ns, is sent to a condenser to compress the primary ions together to form a ~0.6 ns primary ion packet. This increases the current density by a factor of ~30 [16]. With two beam crossovers at Apertures 1 and 2 in the ion column, the primary beam current is further increased, generating more secondary ions per bunched pulse compared to a non-bunched pulse with fewer crossovers. The bunching process unavoidably increases the spread of the beam energy, resulting in high chromatic aberration at the target lens. The spot size in the bunched mode is confined to the micron scale. The beam diameter can be reduced by sacrificing the primary ion current, which leads to a loss in secondary ion intensity. The HCBM has high sensitivity and a very

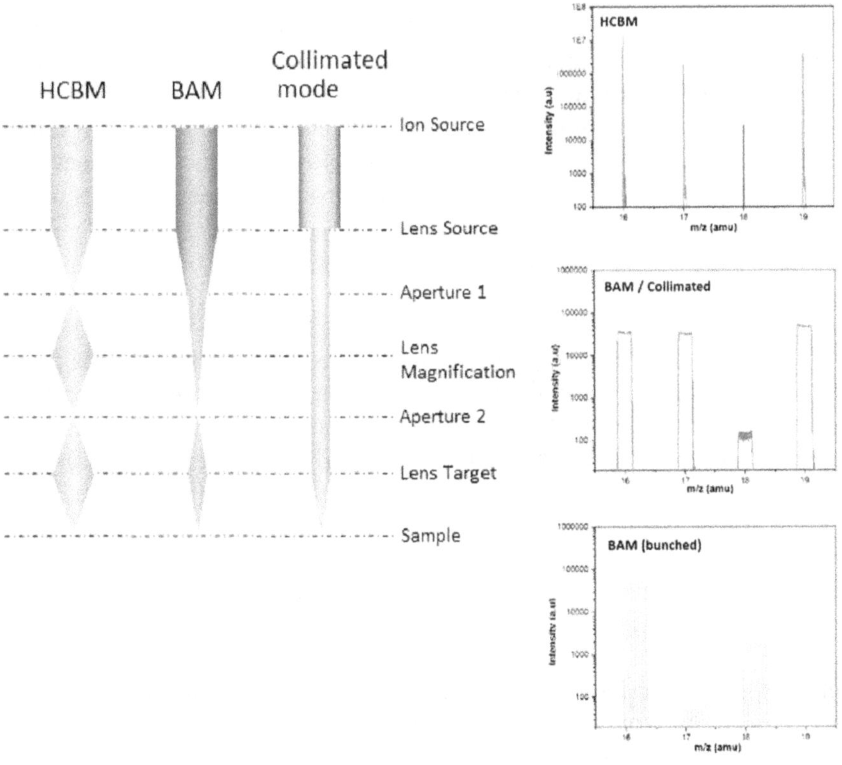

Figure 2.9. Schematic of common beam focusing for different modes of operation in a ToF-SIMS instrument, showing the HCBM, BAM, and collimated mode along with the respective mass spectra obtained in each mode.

high mass resolving power, R, but due to increased chromatic aberrations, it has poor lateral resolving power.

Conversely, the BAM has high lateral resolving power but low mass resolving power (nominal mass), as a much longer pulse ($\sim$100 ns) is used. As there is only a single crossover at Aperture 2, much better beam focus is achieved but sensitivity is compromised, as the beam current intensity is reduced at smaller beam diameters. The BAM is good for ion mapping/imaging but has poor mass resolving power on a nominal mass scale. A second mode is also available with this beam setup, known as 'burst mode'. In this mode, the larger pulse is chopped up into very short pulses of $\sim$1.5 ns, which give a repeating pattern of peaks (four to ten) in the mass spectra (see figure 2.9). As a result of creating this burst mode, the mass resolving power is increased again to around 6000, but sensitivity is lost as the beam current is again sacrificed, leading to reduced secondary ion intensity.

In the collimated mode, there is no beam crossover down the ion column. This mode produces a very well-focused ion beam but at very low currents. The highly focussed beam makes this mode ideal for ion imaging, with a lateral resolving power of $\sim$100 nm. However, the mass resolution is also compromised in this mode and is reduced to the nominal mass (comparable to the mass resolution of the BAM).

Along with the modes outlined above, another mode called *delayed extraction* also exists, which combines the benefits of high mass resolution and imaging modes. The mode is similar to the imaging mode but, as its name suggests, the extraction voltage applied to the ToF analyser that draws in the secondary ions is delayed. It is only applied after a given time. This time delay allows for time-of-flight compensation for the initial secondary ion energy spread. However, this mode does require a higher beam dose to obtain ion images with comparable intensity scales to those of the other modes [17].

2.2.3 Ion beam raster

During experiments, the primary ion beam is moved over the sample in a predefined pattern called a raster. Its main purposes are to produce an analytical map or secondary ion image for analytical beams and to produce a uniform dose of ions for sputtering beams. If the purpose of the SIMS analysis is to obtain a representative spectrum of the surface, then it is also necessary to have a uniform dose of primary ions. For Gaussian-shaped beams, the standard requirement for a uniform dose, with a dose variation of less than 1% away from the edge of the raster, is that the distance between the raster points or lines should be less than half the diameter of the beam. For precise experiments, it is important to understand the relationship between the number of raster points or lines, the size of the raster, and the beam diameter. If the raster side length is greater than half the beam diameter times the number of raster lines or points, then the dose will be nonuniform. For experiments that are set up for a uniform dose, such as depth profiling, it is also important to remember that the edges of the raster area will receive a lower primary ion dose than the centre because of the overlapping dose distributions from each raster position. In depth profiling experiments, these areas should not be included in the analysis region because they occur at different depths, and their inclusion causes poor depth resolution. The width of the avoided border should be at least one and a half times the sum of the sputtering beam diameter and the analytical beam diameter.

The raster sequence typically uses a fast scan axis and a slow scan axis, like old-fashioned cathode ray tube televisions. For continuous sputtering beams, this pattern is achieved by applying sawtooth potentials to the 'x' and 'y' deflection plates. The advantage of the sequential raster pattern is simplicity; however, there are potential disadvantages related to sample charging. Insulating samples usually become positively charged under ion beam irradiation, and this is compensated with a broad beam of low-energy electrons. The electron dose rate is set to balance the sample charging on average across the whole sample, but it is usually insufficient to compensate for instantaneous charging at the beam focus. Sequential raster patterns can cause a local build-up of charge close to the analytical area, which affects secondary ion intensities, as illustrated in the extraction field section. In experiments where both a sputtering beam and an analytical beam are employed in an interlaced fashion (the sputtering proceeds while the time-of-flight analysis from the pulsed analysis is occurring), charging fluctuations may appear as lines of variable intensity caused by charging due to the sputtering beam when the sputtering raster pattern and the analysis raster pattern overlap in time.

To reduce these charge compensation effects, it may be necessary to increase the low-energy electron current, but that risks damaging delicate samples. Some instruments employ a nonsequential or 'random' raster that samples points in the raster pattern in a sequence that avoids near neighbours of recently sampled points. Random raster patterns are useful for simple imaging but do not eliminate all sample charging effects; in some cases, they simply spread the effects uniformly over the image. Poor charge compensation will still be evident as reduced intensity at the edges of an ion image.

2.2.4 Pulsing

Time-of-flight instruments do not scan across a sample's surface with a continuous direct beam. The ion beam is necessarily pulsed across the sample to create the time needed for the emitted secondary ions to be extracted into the analyser and for the ions to fly through the flight tube in order to mass separate. A schematic of the sequential steps is shown in figure 2.10.

During operation, the ion beam pulses across the sample surface in a well-defined raster area. Each pulse has a fixed cycle time, and in each cycle, a sequence of events occurs: the ion beam initially irradiates the sample surface for a very short period of

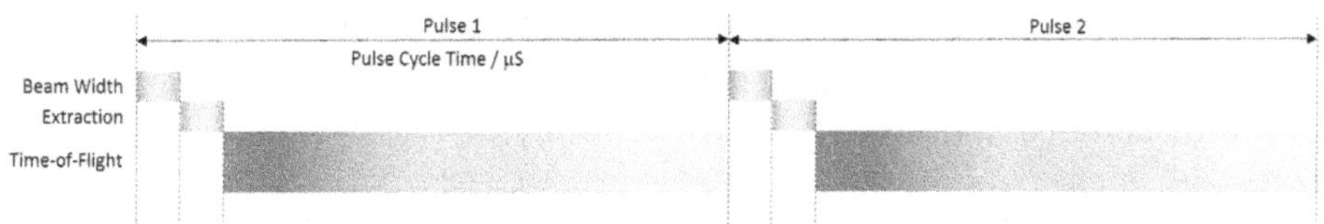

Figure 2.10. Schematic of the ion beam pulsing in a ToF-SIMS instrument, showing the cycle of two pulses.

time (beam width) for between 10 and 25 ns. The beam is then blanked as the sputtered ions are extracted into the analyser column (extraction) for approximately 10 μs. The ions are then left to drift through the flight tube (time-of-flight). The cycle time for each pulse is the time between subsequent ion pulses on the sample surface, and as the cycle time dictates the amount of drift time for the ions in the analyser column, it also dictates the mass range obtained in the mass spectrum. The very short beam width time of the ion beam results in very well-resolved mass spectra, with high mass resolving power, *R*, and high mass resolution (see section 2.4 for a more detailed definition).

2.3 Sputtering and ionisation

2.3.1 Sputtering

When energetic ions impact a solid surface, they interact through collisions with atoms and electrons [18]. Collisions with atoms, known as 'nuclear stopping', are described by coulombic forces between the nuclei of the ion and atoms in the solid. In the ion energy ranges used in SIMS, these 'nuclear' collisions are the dominant process, and kinetic energy is transferred from the ion to an atom in the solid through an elastic collision. If the energetic ion is an atomic ion and either has a smaller mass than the atom with which it collides or a low angle of incidence, then there is some probability that it will travel out of the surface after the first collision. The changes in angular direction and kinetic energy can be directly used to calculate the mass of the atom it collided with. Such backscattered ions are used in ion-scattering spectroscopies such as low-energy ion scattering (LEIS), medium-energy ion scattering (MEIS), and Rutherford backscattering (RBS) to identify and quantify elements at or near the surface of materials. The cross section for atomic collisions, or apparent diameter of atoms to the incoming ion, decreases with increasing kinetic energy. Therefore, this interaction is important for low-energy ions but less important for high-energy ions above 100 keV. 'Electronic stopping', the transfer of energy from the ion to electrons in the solid through the excitation of plasmons and other electronic transitions, results in a continual loss of kinetic energy from the impacting ion. In MEIS and RBS, which utilise high-energy ions, this energy loss may be used to determine the depth at which the ion was backscattered. The excitation of electronic transitions in the path of motion of the ion leads to local heating in the solid.

Both nuclear and electronic stopping lead to atomic primary ions being implanted within the subsurface of the material. For a typical impact energy of 10 keV, the mean implantation depth is less than approximately 10 nm from the surface. The implantation depth decreases for heavy primary ions, high-density materials, and increasing angles of incidence (away from normal incidence), as described in more detail in section 3.3.2. In extended experiments or depth profiles, these implanted ions can be detected and are often used, as in the case of caesium and oxygen, to increase the intensity of other ions through a matrix effect.

The fundamental processes outlined above for an atomic ion colliding with an atom or electrons in the solid apply to the ion after the first collision as well as to an atom recoiling from that collision. In this way, a sequence of binary collisions produces increasing numbers of mobile ions and atoms in the impact zone. This continues until the energy provided by the kinetic energy of the ion has been dispersed among mobile species to the extent that they have insufficient energy to overcome chemical bonds in the solid. The character of this collisional cascade depends upon the nature and energy of the impacting ion and the composition of the material, but can be broadly classified into three groups: linear cascade, thermal spike, and cluster ion impact (see figure 2.11).

Linear cascades occur when the transfer of energy from the primary ion to atoms in the material is inefficient, which is typically the case for single-atom ions such as Ar^+ and Ga^+. Within the excited volume of material, the fraction of atoms that are moving due to the collisional cascade is small. Thermal spikes occur when the transfer of energy from the ion to the material is efficient and results in a large fraction of the atoms in a small region of the material moving with an equivalent temperature higher than 5000 K. This regime is encountered when the primary ion is heavy and usually when it contains more than one atom, for example, in the case of Bi_3^+. When the number of atoms in the impacting ion is as large as many hundreds or thousands of atoms, then the constituent atoms travel too slowly to penetrate far into the target material. The kinetic energy of the projectile acts to compress and heat both the projectile and the target material at the impact zone, and the dynamics are more akin to macroscopic events such as meteor impacts and explosions. In this case, the primary ion initiates a fluid-like motion that excavates a crater in the sample.

In all these processes, there is significant disruption of the material being analysed. This may take the form of damage to the chemical structure of the material or the mixing of atomic species, which changes their distribution. It is important to consider the rate of disruption compared to the rate of removal of material and the rate of secondary ion generation. The sputtering yield characterises the rate of removal and is important both for depth profiling and for secondary ion yields that characterise the strength of the SIMS signal.

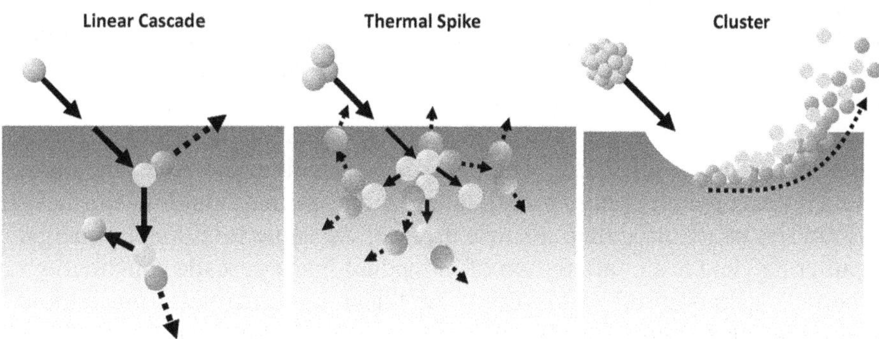

Figure 2.11. Schematic showing the movement of atoms in a solid due to ion beam bombardment: linear cascade, thermal spike, and cluster ion impact.

2.3.2 Sputtering yields

Sputtering yields, Y, are defined as the number of atoms ejected from the material divided by the number of ions that impact the material. This definition is useful in the context of fundamental studies of elemental solids and theory but becomes less helpful when compounds and molecular solids are considered. A more practically useful measure is the mean volume of material sputtered per impact. With knowledge of the 'sputtering yield volume', the mean depth of material removed from the sample is simple to calculate. The sputtered depth is the product of the sputtering yield volume and the ion dose (number of ions per unit area).

In the linear cascade regime, a small fraction of the mobile atoms reaches the surface with sufficient energy to escape. Typically, the sputtering yield for ions with energies of less than 1 keV is close to one and increases with the mass, energy, and incidence angle of the impacting ion. The sputtering yield depends upon the amount of energy deposited in a region close to the surface, which convolves the kinetic energy of the ion with its ability to penetrate the sample. Thus, increasing the kinetic energy of the ion also extends the penetration depth; therefore, the sputtering yield increases less than linearly with kinetic energy. Values may usefully be estimated using software such as SRIM in the linear cascade regime or from relationships published in the literature [19]. Experimental data suggest a modest increase in yield for impact energies between 100 eV and 10 keV, rising by fivefold or so in this range. The sputtering yield reaches a maximum at roughly 100 keV and then starts to fall. The cohesive energy of the material is also important. Materials with strong interatomic bonding, such as Si, Mo, and W, have sputtering yields approximately five times lower than those with weak interatomic bonding, such as Cu, Ag, and Au [20].

Sputtering yields are generally higher at increasing sputtering ion incidence angles [16, 21]. For atomic primary ions, the sputtering yield usually follows a $(\cos\theta)^{-(5/3)}$ relationship, nearly doubling at an incidence angle of 45° and tripling at 60°. At incidence angles higher than 70°, the sputtering yield diminishes due to the increasing propensity of impacting ions to scatter from surface atoms without depositing significant energy into the material.

Heavy primary ions, such as Xe^+, with impact energies higher than 10 keV can have sputtering yields that exceed ten sputtered atoms per incident ion. In these cases, the sputtering yield is higher than that predicted by linear cascade theories because the conditions are within the thermal spike regime. Thermal spikes result in much higher sputtering yields and are characterised by a greater-than-linear relationship between the amount of energy deposited in the near-surface region and the sputtering yield [22, 23]. Thermal spikes usually occur when the impacting ion is a cluster ion containing more than one atom. In this situation, the individual atoms are less penetrative than a single atom of the same total kinetic energy, and the sputtering yield is a combination of individual linear cascade contributions and the thermal 'evaporation' yield. Both the linear and thermal contributions are considerably larger than simple linear cascade predictions. The linear cascade contribution to the sputtering yield is larger than that for a single atomic ion of the same mass and energy because sputtering yields for single atoms are only weakly dependent on impact energy, so increasing the number of atoms in the primary ion increases the yield to a greater extent than the decreasing energy per impacting atom reduces the yield. The thermal yield, Y_{th}, is approximately proportional to the square of the impact energy for penetrative ions. Thus, the sputtering yield for cluster ions can easily reach 100–1000 sputtered atoms per incident ion.

For large clusters, typically with many more than ten atoms, the thermal spike predictions overestimate sputtering yields. In these cases, the amount of energy deposited in the sample is largely confined to the surface region, and the energy of the ejected material cannot be larger than the impact energy. Such situations, where the sputtering yield becomes proportional to the impact energy, constitute the cluster sputtering regime [24] (see figure 2.12).

Sputtering yields from cluster ions such as carbon (C_{60}) and argon (Ar_n) clusters can be very large, particularly for organic materials. Yields are typically larger than 1000 sputtered atoms (approximately 10 nm^3) per incident ion but also depend upon the nature of the material. The sputtering yield can be characterised by the relationship between (*Y*/*n*) and (*E*/*n*), where experimental sputtering yields for specific materials lie close to a single curve (see figure 2.13) [25–27]. When (*E*/*n*) is large, the relationship is linear and thus the yield is proportional to the impact energy of the ion irrespective of the number of atoms in the cluster ion. For small values of (*E*/*n*), i.e. for large clusters and small impact energies, the sputtering yield is smaller than expected from this linear relationship. The transition in behaviour for argon cluster ions, which are the most studied type of cluster ion, occurs in the range of 1–10 eV/atom for organic materials and 10–100 eV/atom for inorganic materials. It is interesting to note that this transition occurs when the velocity of the cluster is

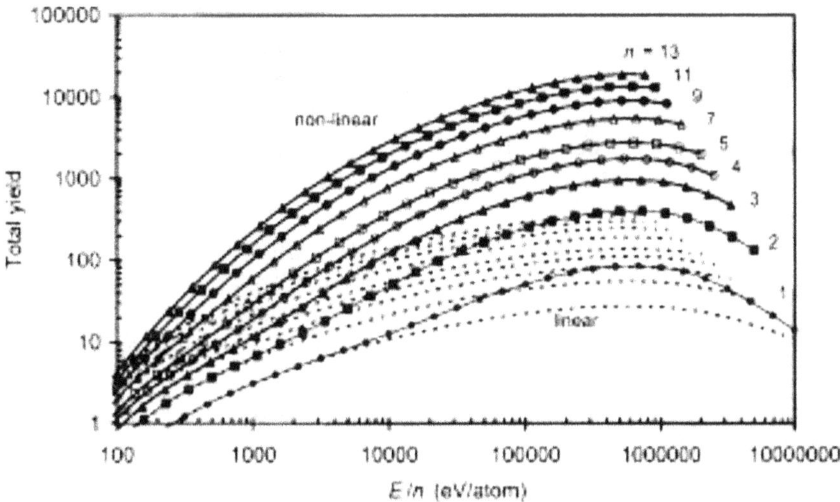

Figure 2.12. Sputtering yields for gold sputtered by Au_n^+ clusters compared to the results predicted by linear cascade theory. Reprinted from [24], with the permission of AIP Publishing and the American Vacuum Society.

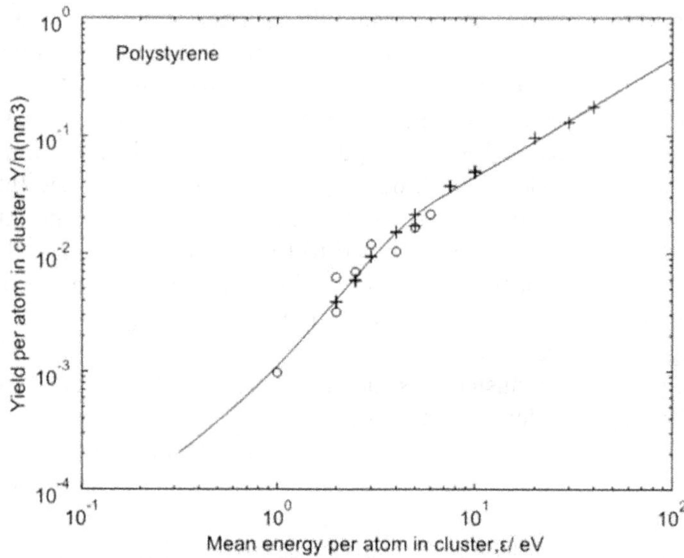

Figure 2.13. Measurements of total argon cluster ion sputter yield from (polystyrene) with a descriptive curve. Reprinted with permission from [26]. Copyright (2013) American Chemical Society.

only a few times larger than the speed of sound in the material, implying that an elastic response to the impact can occur and that such a response may act to dissipate the impact energy.

For organic materials, the sputtering yield volumes for argon clusters are approximately 5–10 nm^3 keV^{-1} per incident ion in the regime where the yield is proportional to the impact energy and the impact angle is 45°. These yields are influenced by the cohesive energy of the solid, with molecular materials having a higher sputtering yield than polymers. For some materials, it has been noted that sputtering yields increase with the sample temperature [28].

The angle of impact is also important for organic materials, particularly when (E/n) is less than 10 eV/atom. In this case, very low sputtering yields are found at normal incidence, and the maximum sputtering yield is obtained at angles of incidence close to 45°. Compared to the sputtering yield at normal incidence, this maximum becomes more pronounced for values of (E/n) less than 10 eV/atom (see figure 2.14) [29]. Molecular dynamics simulations [30] support the conclusion that sputtering yields are highest close to a 45° impact angle and suggest that the likely cause of this increased sputtering yield is an induced lateral motion of material in the impact zone. Further evidence of this lateral motion is found in the asymmetric distribution of ejected material during sputtering. For off-normal impacts, even at angles as small as 15°, most of the ejected material retains the lateral motion of the impacting ion and is ejected in the downrange direction, with virtually no ejection in the opposite direction [31]. The emission angle of the majority of ejected material is close to 55°, with little dependence on the argon cluster impact angle (see figure 2.15).

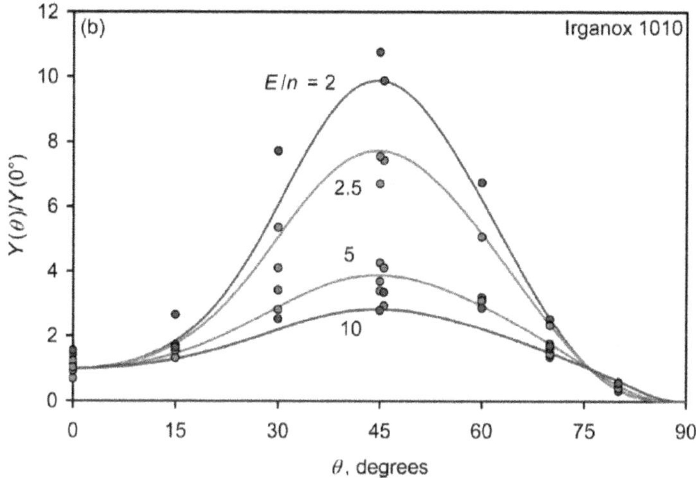

Figure 2.14. Irganox 1010 sputtering yield dependence on incident angle for different Ar GCIB clusters. Reprinted with permission from [29]. Copyright (2015) American Chemical Society.

2.3.3 Ionisation

For SIMS to work as a technique, it is important that ionisation occurs during or after the sputtering event. Ionisation in the impact zone is a consequence of the extreme temperatures and pressures temporarily generated in the impact zone. The energies required to ionise atoms and molecules in condensed phases are similar to the energies required to break chemical bonds. It is likely that a significant proportion of atoms and molecules in the impact zone are ionised immediately after the primary ion impact. However, it is currently not possible to predict with any confidence what proportion of the ejected material will be ionised, nor which of the ejected species will be ionised. Ionisation is complicated by the exchange of charge between species in the impact zone and neutralisation processes that occur during ejection. Neutralisation may occur through interactions between secondary species of opposite charge but is more likely to occur between the secondary ions and the surface. For ions close to pristine metallic and semiconductor surfaces, resonant and Auger neutralisation processes have been studied in detail.

As a result of these processes, only a small fraction of particles ejected after a sputtering event are ionised. In most circumstances, the fraction of species that are ionised is smaller than one in a thousand of the number ejected. Comparing this typical ionisation probability to the typical sputtering yields given in section 2.3.2 demonstrates that most primary ion impact events produce no secondary ions. The exception is cluster ion beam sputtering, where the sputtering yield is high enough that, on average, a few secondary ions are generated during each impact event.

The charge exchange processes depend upon both the secondary ion that is being detected and the material close to it during the sputtering event. This surrounding material may enhance or suppress the intensity of a particular secondary ion, and this dependency upon the surrounding material is known as the 'matrix effect'. Secondary ions that have a low formation energy have high relative intensities, while

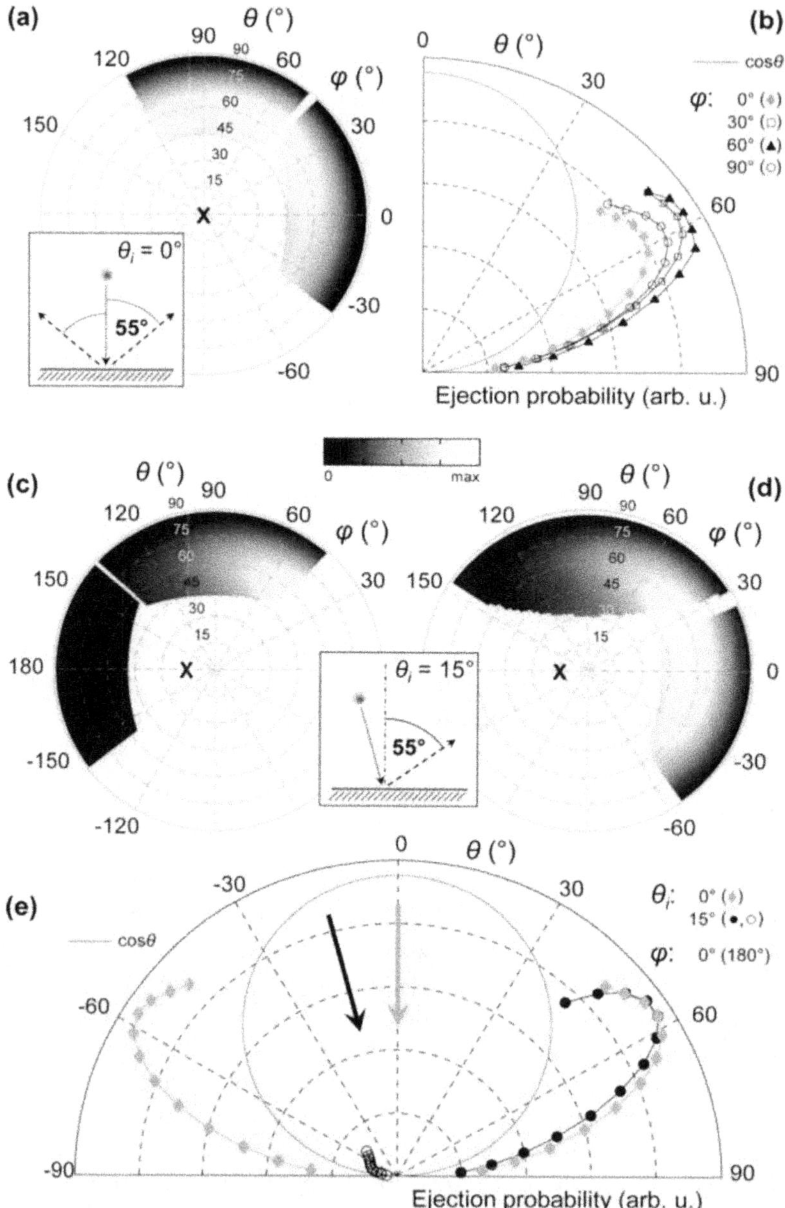

Figure 2.15. Angular distribution probabilities of Irganox 1010 ejected after Ar GCIB impacts at different incident angles. The ejection is highly directional, even for impacts 15° away from normal incidence. Reprinted with permission from [31]. Copyright (2016) American Chemical Society.

those with high formation energies have weaker intensities and may even be unobservable. Because of the matrix effect, the intensities of secondary ions cannot be transformed into compositional information without additional knowledge. For dilute analytes in a matrix with an otherwise constant composition, linearity between analyte concentration and secondary ion intensity is expected. In this case, a constant sensitivity factor can be applied to transform the analyte secondary ion intensity into a concentration (see section 4.1).

If the matrix material changes, then the ionisation probability for the secondary ion also changes, as does the sensitivity factor. In this case, even if there is a uniform concentration of the analyte in the sample, the secondary ion intensity may change in an image or depth profile as the concentrations of the different phases change. For samples that have more than one phase in the analysis area, reliable quantitative analysis requires significant effort. To perform a quantitative measurement of concentration, it is therefore necessary to have a set of reference materials where the concentration of the analyte is known and contained within the same material being analysed. Given enough different analyte concentrations covering the concentration range required, a calibration curve can be established.

2.3.4 Extraction fields

Many types of SIMS instruments use an extraction field to accelerate secondary ions away from the surface into the analyser. The kinetic energy of the secondary ions entering the analyser is usually controlled by the extraction field, and transport through the spectrometer to the detector relies upon the ions having similar kinetic energies. Therefore, the extraction voltage is set to produce kinetic energies much higher than the typical ~2 eV range of secondary ion emission energies. For flat, conductive samples, the extraction voltage is a direct indication of the kinetic energy of the secondary ions entering the mass analyser. However, for flat, insulating samples, this is not the case. The surface potential of an insulating sample depends upon the charge compensation settings used, as well as the thickness and dielectric constant of the sample above a conductive support. For such samples, the spectrometer settings, such as the reflector potential in reflection time-of-flight analysers, require adjustment to optimise mass resolution.

In the case of topographic samples, the interaction of the sample with the extraction field should be considered. This is particularly important if meaningful and complete secondary ion images are desired. If the samples are conductive, then the effects of topography can be severe due to lateral deflection of ions in the fields close to the sample surface. This is exemplified by the image of a 125 μm diameter gold wire on a conductive substrate [32] shown in figure 2.16. The gold wire is visible

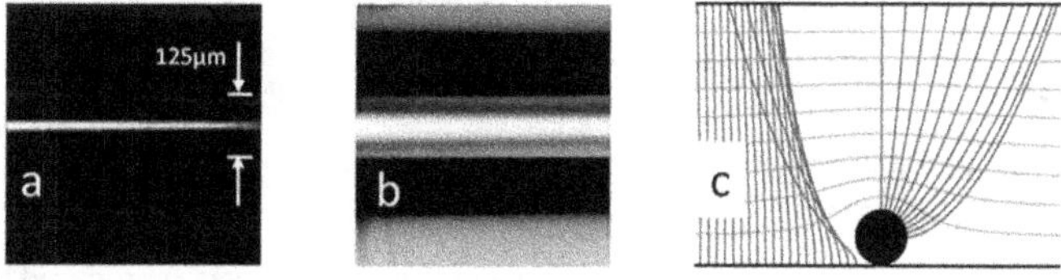

Figure 2.16. (a) Total secondary ion image on a linear colour scale of a gold wire on a silicon wafer. The diameter of the wire is marked on the figure. (b) The same image on a logarithmic colour scale. (c) Schematic of the electric field equipotential lines (green) and the resulting secondary ion trajectories (blue). Reprinted with permission from [32], with permission from Springer Nature.

on the linear intensity scale in panel (a) but is much narrower than expected. The logarithmic intensity scale in (b) shows that ions are mainly detected from the apex of the wire, but not from the sides. Additionally, there is a shadowed area of the substrate close to the wire that extends further than the wire diameter on either side. The schematic in (c) demonstrates how the conductive wire distorts the extraction field and causes lateral acceleration of the secondary ions both from the wire (right-hand side) and from the substrate (left-hand side). Although some secondary ions appear to be visible from the sides of the wire in panel (b), these predominantly arise from primary ions that are scattered from the wire and then impact the substrate outside the shadow region. The size of the shadow regions may be reduced by decreasing the extraction potential or, in a time-of-flight instrument, delaying the extraction pulse until the secondary ions have moved some distance from the surface. Both approaches can compromise mass resolution.

Similar effects occur when the surface is differentially charged, and this commonly happens when insulating and conductive regions of a sample are close together. Figures 2.17(a) and (b) illustrate the effect of a differentially charged boundary on the extraction field and secondary ion trajectories: the deflection of secondary ions produces a shadow region at the boundary. Experimental results from an earthed indium tin oxide conductor with an insulating photoresist are shown in panel (c). The electron flood gun potential and current have a strong influence on the surface charge of the insulator and the size of the shadow region. Careful control of flood gun settings, as well as the measures outlined for topographic conductors, may reduce the size of the boundary. For this type of sample, it may also help to disconnect the conductor from earth and allow it to float to the electron source potential.

Similar shadows become visible at the edges of secondary ion images of insulating samples if the electron flood current is too low to compensate for the charging induced by the ion beam. The ion-irradiated area charges with respect to the surrounding areas that are close to the electron source potential, and this creates a

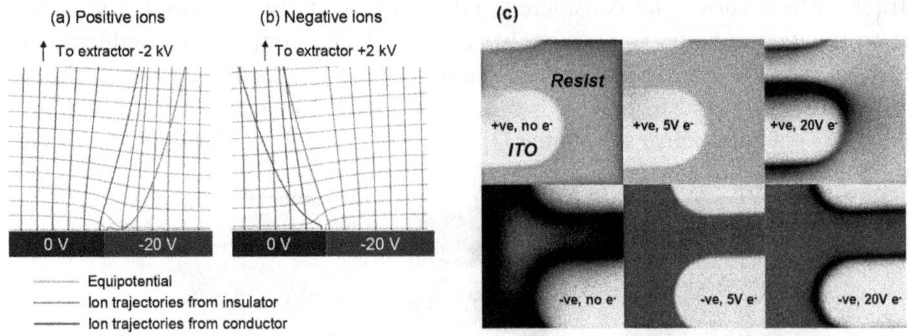

Figure 2.17. (a) and (b) Schematic of the electric field equipotential lines (green) produced by a differentially charged surface and the resulting secondary ion trajectories originating in the earthed region (blue) and the charged region (red). (c) Experimental results from the boundary of an earthed indium tin oxide conductor and insulating photoresist, showing the effect of electron flood gun settings. The top row contains positive ion images, while the bottom row consists of negative ion images. (Figure courtesy of Alex Shard, NPL.)

similar boundary and shadow. These edge effects are evident in figure 2.15(c) when the electron flood gun is off, and the positive charge induced by primary ion irradiation has a significant effect on the negative secondary ion image.

2.4 Mass analysis

Once the secondary ions have been generated by the primary ion beam rastering across the sample surface, they need to be separated according to their mass-to-charge ratio (m/z). There are several commonly used mass analysers: quadrupole, magnetic sector, time-of-flight, mass spectrometry/mass spectrometry (MS/MS), and Orbitrap. Each of these analysers uses different physical principles to separate the secondary ions; consequently, each has a different upper mass limit, mass resolving power, mass accuracy, and mass detection [33]. Regarding the mass analysers described in the following section, namely quadrupole, magnetic sector, time-of-flight, and Orbitrap analysers, some general specifications are shown in table 2.1.

For all mass analysers, the ability to separate peaks in mass spectra is paramount. It is therefore important for the mass resolution to be well defined [34]. In mass spectrometry, two different parameters, namely mass resolving power and mass resolution, are interchangeably used, which can lead to confusion, and there is no consensus within the field of SIMS as to which parameter is the gold standard. Both resolving power and resolution are, therefore, described below.

2.4.1 Mass resolving power, *R*

Mass resolving power, R, is an important parameter, as it describes the ability of an analyser to accurately measure a species of interest. During the collection of secondary ions, mass interferences can occur if different ions have the same nominal mass, M. For example, ^{27}Al and $^{12}C_2{}^1H_3$ or ^{31}P and $^{30}Si^1H$ have nominal masses of 27 and 31, respectively. However, the exact atomic masses are not the same due to differences in nuclear binding energies. The mass resolving power is calculated by dividing the nominal mass of interest by the difference between the precise masses of the interfering species (see equation (2.2)). For example, in the case of mass 27, the ΔM between ^{27}Al and $^{12}C_2{}^1H_3$ is 0.042. Therefore, the mass resolving power is $R = 27/0.042 = 643$:

Table 2.1. General specifications of mass analysers commonly used in (SIMS).

Analyser	Mass resolving power, R	Extraction field	Transmission factor (F)	Mass detection	Mass range
Quadrupole	1000	$\sim$10 V mm^{-1}	10^{-3}	Sequence	4000
Magnetic sector	100 000	$\sim 10^3$ V mm^{-1}	10^{-2}	Sequence	20 000
Time of flight	>10 000	$\sim 10^3$ V mm^{-1}	10^{-1}	Parallel	1 000 000

Table 2.2. Some common interfering species.

Nominal mass	Interfering species				Mass resolving power, R	
		Mass		Mass	ΔM	$M/\Delta M$
14	N	14.0031	CH_2	14.0157	0.013	1115
16	O	15.9949	CH_4	16.0313	0.036	439
27	Al	26.9815	C_2H_3	27.0235	0.042	643
31	P	30.9738	^{30}SiH	30.9816	0.008	3959
32	S	31.9721	O_2	31.9898	0.018	1808
39	K	38.9637	NaO	38.9847	0.021	1857

$$R = \frac{M}{\Delta M}. \tag{2.2}$$

A small selection of some simple and common mass interferences and the mass resolving powers needed to separate each species are shown in table 2.2.

The accuracy of peak assignment is essential in SIMS, but the precision that is needed is also dependent on the samples being analysed. For the mass interference highlighted in table 2.2, a quadrupole-based SIMS instrument would be able to accurately separate and measure the interfering species (table 2.1 also lists R for the three most common mass analysers). For biological materials, where distinguishing proteins, lipids, or enzymes relies on accurately labelling fragments based solely on different combinations of C, H, O, N, or S, for example, much greater mass resolving power is needed to avoid the inevitable mass interference and peak overlap. To this end, time-of-flight and Orbitrap-based SIMS instruments are commonly used. In the fields of geological research and the accurate measurement of isotopes, only the extremely high mass resolution obtained using larger magnetic sector instruments is suitable.

2.4.2 Mass resolution

Another common definition of mass resolution, and a way to ensure the consistency and performance of an analyser, is to also measure the full width at half maximum (FWHM) of a specific peak and calculate the mass resolution $m/\Delta m$ (please note that I have used lower-case m to differentiate between this and the mass resolving power described previously), as shown in figure 2.18.

In this instance, in the case of many time-of-flight instruments, the H or Si ion peak is typically selected. The mass peak position (m) is measured, along with the width of the peak (Δm) at 50% of its peak maximum. The resolution of this peak can therefore be monitored for different analyses, and if the peak begins to broaden, it can be an indication that the mass analyser needs to be recalibrated, or ion beam pulse widths adjusted; alternatively, the peak may have been broadened due to surface roughness.

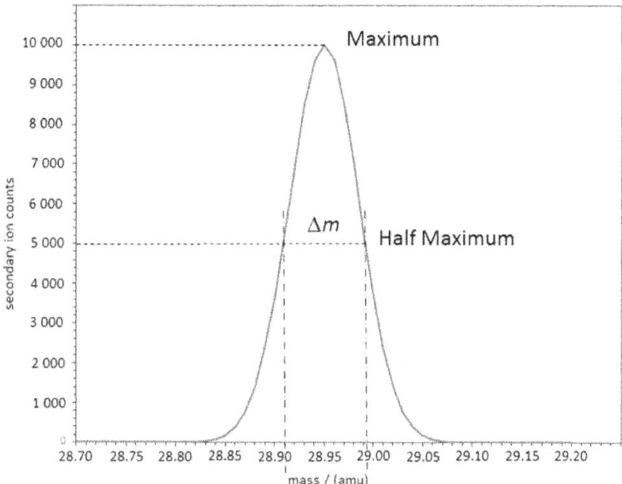

Figure 2.18. Example of measuring the mass resolution, Δm, at the FWHM of a Si^+ ion peak.

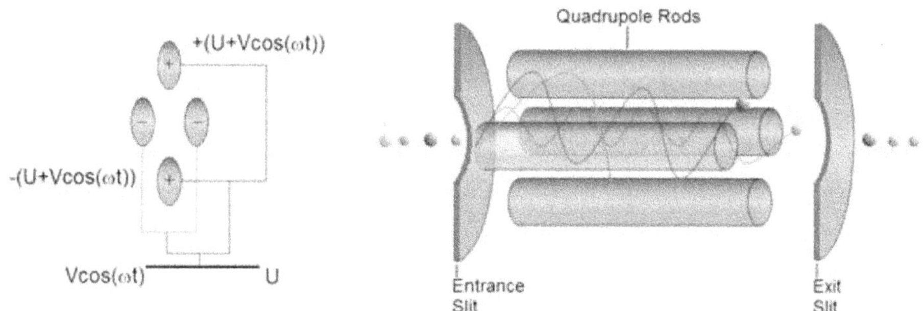

Figure 2.19. Schematic of a quadrupole mass analyser. (a) Radial view of voltages applied to the rods and (b) secondary ions travelling through the quadrupole, showing stable (red and purple) and unstable (blue) ion trajectories.

2.4.3 Quadrupole analysers

A quadrupole mass analyser, as the name suggests, is composed of four circular poles or rods. The four parallel rods have both DC and RF (radio frequency) voltages applied to them. The DC voltage is fixed, and the RF voltage is alternating, while the RF voltage for one pair of rods is 180° out of phase with the voltage on the other pair of rods. The potential on one pair of rods is therefore $+(U + V\cos(\omega t))$, and the other pair of rods has the voltage $-(U + V\cos(\omega t))$, where U is the fixed DC potential and $V\cos(\omega t)$ is the RF voltage that varies with frequency, ω, and time, t (see figure 2.19(a)).

As the ions are sputtered from a target material and injected into the analyser, their trajectory through the quadrupole is complex, and a simple schematic is shown in figure 2.19(b). Simply put, the ions pass through the field-free region along the central axis of the rods, oscillating between them, with trajectories dependent on the mass-to-charge (*m*/*z*) ratio of the ions. For a given DC and RF voltage combination, only ions with a specific *m*/*z* ratio undergo stable oscillation and pass through to the detector. All other masses with unstable oscillation hit the rods and are not detected.

The larger the rods, the better the performance in terms of secondary ion transmission and mass resolving power. Only selected mass channels can be collected using a quadrupole analyser; however, the dynamic range of detection can be six or seven orders of magnitude.

2.4.4 Magnetic analysers

Secondary ions generated by the bombarding primary ion beam have a wide range of energies. A magnetic analyser alone cannot effectively separate these ions, so it is combined with an electrostatic analyser (ESA). The ESA helps to reduce aberrations and attain high mass resolution. This type of instrument is often known as a double-focussing instrument (see figure 2.20).

Initially, the secondary ions enter the ESA, where the ions are filtered according to their kinetic energy (i.e. in a mass-independent manner). Low-energy ions are deflected more than high-energy ions as they pass through the electric field. A movable energy window is positioned to select a small portion of the sputtered ions. By shifting the window toward the higher-energy ions, molecular species are suppressed. An electrostatic lens then aligns the energy-filtered ions so that they enter the magnetic analyser.

Mass separation (*m*/*z*) takes place as the ions pass through the magnetic field. A force at right angles to both the direction of motion and the magnetic field acts on the ions, and the mass-to-charge ratio (*m*/*z*) separation is determined by the following equation:

$$\frac{m}{z} = \frac{B^2}{2V} \times r^2, \tag{2.3}$$

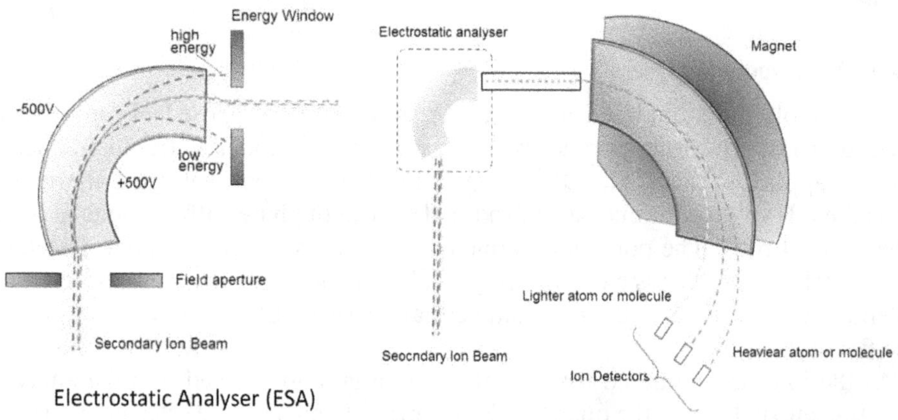

Figure 2.20. Schematic of a magnetic sector mass analyser and an ESA.

where B is the magnetic field, V is the accelerating voltage of the secondary ions, and r is the radius of curvature with which the ion passes through the magnetic field.

Magnetic sector analysers are typically large and expensive but provide very high mass separation (R) (10 000–25 000) and high detection sensitivities. There can be limitations on the number of ions collected at any one time, and this must also be preselected.

2.4.5 Time-of-flight analysers

A time-of-flight (ToF) mass analyser is the simplest of mass analysers, which is probably why it is also one of the most well-known types of mass analysers and is found in many branches of mass spectrometry. In order for it to be used in SIMS, the secondary ions ejected from the sample surface have to be produced in a series of extremely short pulses, meaning that the ion beam itself is rastered over the chosen area in a pulsing fashion, and the ion beam rests on the sample for a very short period of time (of the order of 10–20 ns). Once the secondary ions are ejected from the target surface, they are accelerated into the ToF analyser's flight tube through an extraction plate held at a fixed potential, V, typically between 2 and 8 KeV (see figure 2.21). The potential energy E_p of the particles is thus converted into kinetic energy E_{KE}. E_p and E_{KE} are equal such that:

$$E_p = E_{\mathrm{KE}}$$

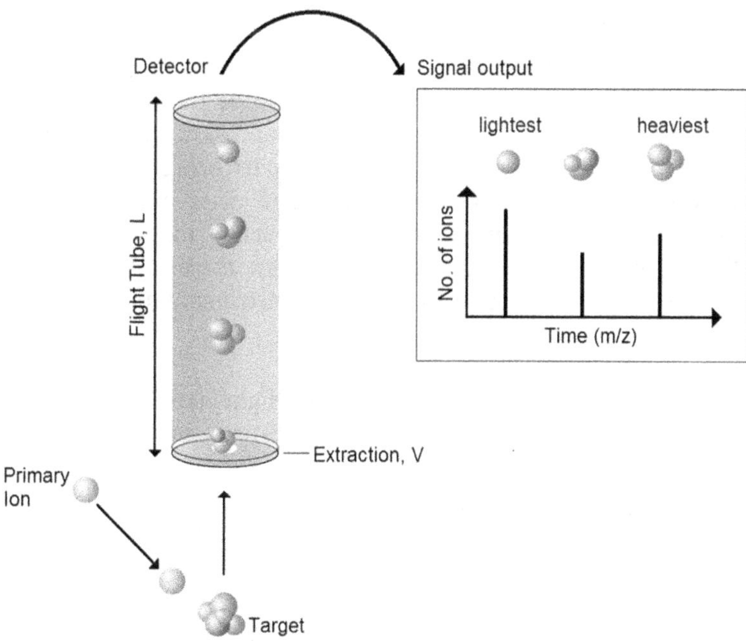

Figure 2.21. Schematic showing the time of flight of sputtered secondary ions through the flight tube. The lightest secondary ion arrives first at the detector, followed by the heavier secondary ions.

$$zV = \frac{1}{2}mv^2, \tag{2.4}$$

where z is the charge of the particle, V is the accelerating potential, m is the mass of the particle, and v is its velocity. The velocity of the ions after acceleration is constant as the particles fly through a field-free tube. Their velocity is determined by the length of the flight tube, L, and the time they take to arrive at the detection system; therefore,

$$v = \frac{L}{t}. \tag{2.5}$$

Substituting equation (2.5) into equation (2.4) allows the mass-to-charge ratio of the sputtered secondary ions arriving at the detection system to be calculated as follows:

$$\frac{m}{z} = \frac{2Vt^2}{L^2}. \tag{2.6}$$

The lightest ions travel the fastest and arrive at the detector first. The heavier ions move more slowly and arrive at later intervals. The arrival times of the ions at the detector are dependent on the ion mass; thus, using equation (2.6), the arrival times can be transformed into the respective m/z ratio, generating a mass spectrum from each pulse of the ion beam.

At higher masses, resolution becomes more difficult as flight times are longer, and not all the ions with the same m/z ratio reach their ideal ToF velocities. To remedy this problem, a series of high-voltage ring electrodes is placed at the end of the analyser; this is known as a reflectron.

Reflectrons improve the resolution at higher masses by narrowing the range of flight times for a single m/z value. Faster ions travel further into the reflectron compared to slower-moving ions. Thus, both fast and slow ions with the same m/z value reach the detector at the same time, thus narrowing the bandwidth of the output signals. For a commonly used reflectron-ToF mass analyser, the upper mass limit that can be recorded has an m/z of 10 000. Typically, mass resolution may be as good as 20 000 with an accuracy of 10 ppm, but these limits of detection are dependent on many factors related to individual instrumentation.

2.4.5.1 Mass calibration

Along with the simplicity of design, the time-of-flight mass analyser also has the advantage that it is self-calibrating, as two very distinct mass spectra are obtained for positive or negative secondary ions. Internal mass calibration can consequently be based on the light mass fragment ions that are always present in a mass spectrum. The low mass fragment patterns for the positive and negative secondary ions are shown in figures 2.22(a) and (b) respectively. Both spectra show an initial peak, followed by a collection of peaks with varying intensity. The first peak that is observed at (nominally) 1 amu is that of hydrogen. There then follows a 'packet' of peaks corresponding to carbon and three hydrocarbon molecular fragments: CH, CH_2, and CH_3. The intensities of each of these peaks differ depending on the

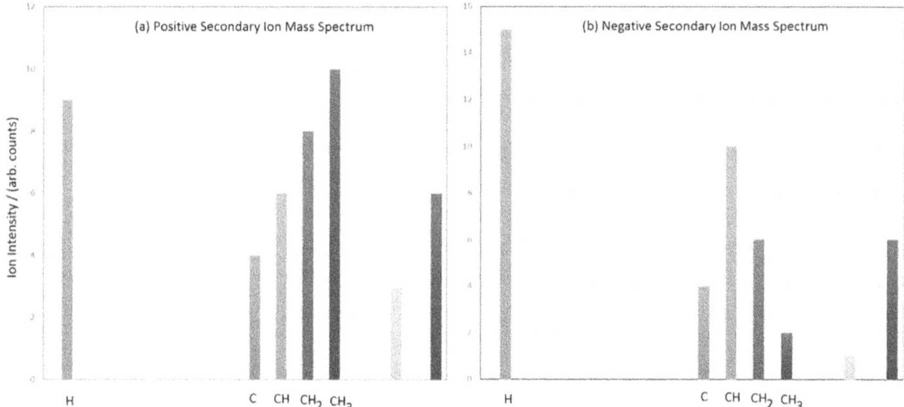

Figure 2.22. Schematic of the lower mass fragment ion peaks used for calibration of the *x*-axis: (a) positive secondary ions and (b) negative secondary ions.

Table 2.3. Example of ion fragments used for calibrating the *m/z x*-axis of mass spectra.

Ion polarity	Sequence	Example fragments
Positive	C_xH_{2x-1}	C_2H_3, C_3H_5, C_4H_7, C_5H_9
Negative	C_xH_2	C_2H, C_3II, C_4H, C_5H

polarity of the secondary ions being detected. In the positive ion mode, the peak of CH_3^+ is the most intense, with the preceding peaks dropping in intensity. In the negative ion mode, the peak of CH^- is the most intense, with that of the CH_3^- ion now being the least intense and in some instances not visible at all. The ability to recognise these two patterns at the lower mass end of the positive and negative ion mass spectra allows for the calibration of the *x*-axis from time to *m/z*.

Calibration needs to be continued along the *x*-axis. Again, simple hydrocarbons are used and again differ between the two polarities. Table 2.3 highlights some fragment peak sequences used to continue the calibration. For positive mass spectra, the formula $C_xH_{2x-1}^+$ can be used, with peak intensities typically higher for odd H ion contents. For negative secondary ion mass spectra, the C_x^- and/or C_xH^- series may be used, along with the very electronegative elemental ions such as F^- and Cl^-. As the precision of time-to-digital converters is very high (these devices have very good linearity over the full range), the relationship between $(m/z)^{1/2}$ and the ToF remains linear over the whole mass range. Ideally, the *m/z*-axis should be calibrated up to at least 50% amu of the heaviest peak of interest.

2.4.6 Orbitraps

Fourier transform mass spectrometers operate by using ion traps to confine ions and set them upon stable trajectories within the trap. The period of oscillation within the trap depends on the mass-to-charge ratio of the ion. The circulating ions induce a current that can be converted from the time domain to the frequency domain using a Fourier transform. The primary advantage of Fourier transform instruments is that they have very high mass resolving powers, which often increases the certainty of identification and enables the separation of isobaric interferences. The mass resolving power increases with the circulation time of the ions, and spectral acquisition times of greater than 100 ms are generally required. This is slow compared to other mass spectrometers and can make experiments lengthy.

Fourier transform ion cyclotron resonance (FTICR) mass spectrometers operate by using a strong magnetic field to induce a circular motion in ions, the period of which is inversely proportional to the mass-to-charge ratio of the ion. FTICR analysis has been used in SIMS to study inorganic [35] and biological [36] materials.

Orbitrap mass spectrometers [37] use a modified electrostatic Kingdon trap [38], where the ions circle a central spindle. The shape of the central spindle is modified so that the ions oscillate axially between one half of the trap and the other, with a frequency that is proportional to the inverse square root of the mass-to-charge ratio. The combination of Orbitrap analysis and SIMS has been shown to be useful for biological analysis [39–41] and is starting to be used in other technological areas [42].

2.5 Secondary ion detection

During the SIMS process, the number of ejected secondary ions is very low, and the majority of species ejected are neutral and have no charge. As the fraction of sputtered charged ions is very low, usually less than 1% of the sputtered particles [43], the detected secondary ion signal is amplified to improve the counting statistics. Modern SIMS instrumentation may have more than one detector. Each manufacturer has its preferred design, and some instruments can have as many as 12 detectors (www.cameca.com).

2.5.1 Electron multipliers

The most sensitive detector is the electron multiplier, which has a high signal gain and low noise. The background count rate is typically less than 0.01 counts per second (noise can arise from stray ions and cosmic rays), but it must be protected from intense ion beams, as this can lead to a decline and eventual failure in its operation. Electron multipliers are used in different geometries, with the most common being the discrete dynode, the continuous dynode, and microchannel plates (MCPs; see the figures below for their configurations). To collect both positive and negative ions, the casing of the electron multiplier is set to the ground potential. The pulses produced in an electron multiplier by the striking ions are subsequently amplified, thus improving the counting statistics.

The time taken by the detector electronics—amplifiers and a discriminator—to process the pulse is referred to as the dead time, τ. To minimise the dead time, count rates are often limited to $\sim 10^6$ counts per second, and the true count rate, n, may be calculated from the measured count rate, n_o, using the following equation:

$$n = \frac{n_o}{1 - \tau n_o}. \tag{2.7}$$

For a typical measured count rate, n_o, of 5×10^5 counts per second and a dead time of 20 ns, the true count rate, n, would be 1% higher.

Pulse-counting detectors follow Poisson statistics, which require each ion to arrive independently of one another. Over a period of counting time, n counts are detected. The standard deviation of the count measurement is therefore given by:

$$SD = \sqrt{n}. \tag{2.8}$$

2.5.1.1 Discrete dynode

In this setup, a series of dynodes is connected by a string of resistors. Ions are drawn into the first dynode by a large negative voltage, and the final dynode is held at ground potential (see figure 2.23).

Along the chain of dynodes, there is a gradual voltage drop from front to back. Typically, the voltage difference between each successive dynode is 100–1200 V. When an ion strikes the first dynode, it produces secondary electrons. The secondary electrons are accelerated into the next dynode, where each electron produces more secondary electrons. An avalanche of secondary electrons ensues as the process continues down the dynode chain, and the electron count increases exponentially. The magnitude of the final pulse is dependent on the number of dynodes, the accelerating voltage between them, the impact energy, and the type of charged particle striking the dynodes.

2.5.1.2 Continuous dynode (Channeltron)

An electron multiplier can also be formed using one continuous dynode, also known as a Channeltron, as shown in the schematic of figure 2.24. The difference between the two geometries is that the continuous dynode has enough resistance to create a gradual voltage drop from the front of the dynode to the back. The operation of the continuous dynode is analogous to that of the discrete dynode: a high voltage at the front of the dynode accelerates ions into the dynode, and electrons are generated upon striking the detector wall. These are accelerated along the dynode to the back of the detector due to the potential drop, creating a cascade of electrons each time

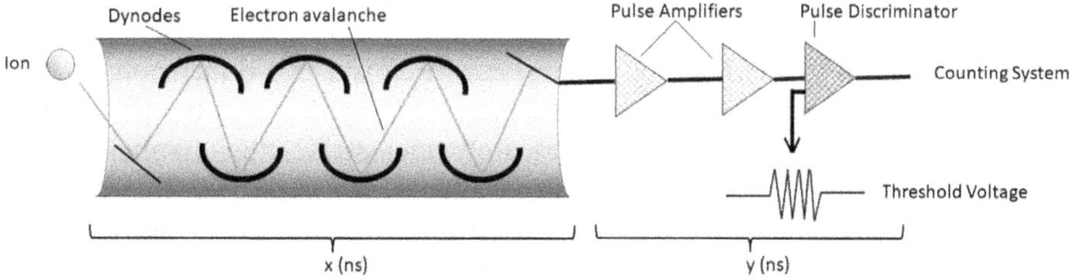

Figure 2.23. Schematic of a discrete dynode detector.

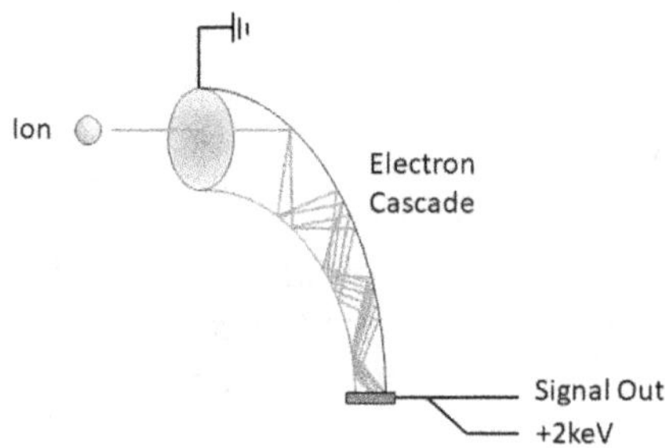

Figure 2.24. Schematic of the ion path and resulting electron cascade created in a continuous dynode (or Channeltron).

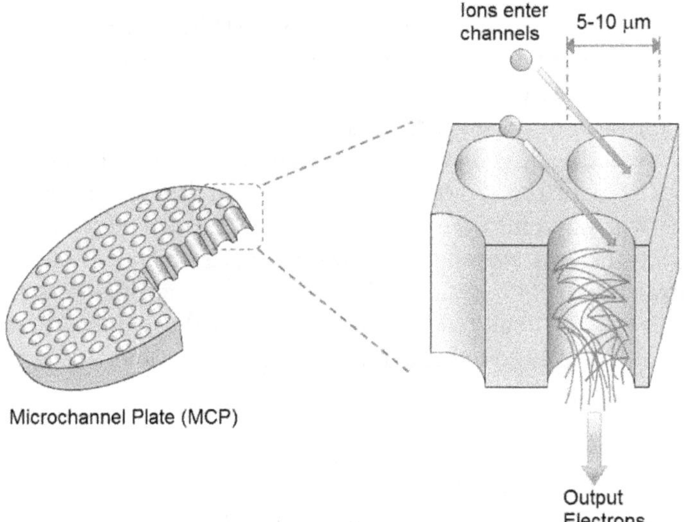

Figure 2.25. Schematic of a microchannel plate used in the detection of secondary ions.

they strike the detector walls. An exponential increase in electron counts is again generated along the length of the detector.

2.5.2 Microchannel plates

Ion image detectors are dependent on MCPs. MCPs combine several unique properties such as high gain, high spatial resolution, and high temporal resolution, making them useful for many applications. An MCP is a two-dimensional array of very small glass channels (~5–10 μm) that are fused together and then sliced into a plate (see figure 2.25). A single ion enters a channel, strikes the channel wall, and emits an electron. As there are many channels in the plate, the process is continuously repeated, creating a cascade of several thousand electrons that emerge from the back of the plate. More than one MCP may be operated in series, leading to an amplification of a single event into an output of $\sim 10^8$ electrons.

2.6 Ultrahigh vacuum

Along with the development of ion beam sources and faster electronics for better counting statistics, many of the advances and improvements in sensitivity in mass spectrometry have occurred due to enhancements in the vacuum environment. Although this is not strictly a part of the analytical technique, the importance of the vacuum cannot and should not be underestimated, as it is a key ingredient needed for any high-sensitivity surface science experiments.

The vacuum in a modern ToF-SIMS analysis chamber can be as low as 10^{-10} mbar. This is obtained through an arrangement of roughing pumps, turbomolecular pumps, ion pumps, and sublimation gettering pumps. Such low vacuum pressures are essential for two reasons: first, to maintain a surface that stays clean for the duration of the surface science experiment, and second, to maximise the inelastic mean free path taken by the generated secondary ions in traversing the analysis chamber into the analyser without encountering losses due to gas-phase scattering phenomena.

Using the kinetic theory of gases, we can make an estimate of the time taken for a surface to be covered by a monolayer of adsorbate (e.g. dry air) and therefore how long a surface stays clean.

The incident flux of gas molecules arriving at a surface is given by equation (2.9):

$$F = \frac{P}{\sqrt{2\pi M k_{\mathrm{B}} T}}, \tag{2.9}$$

where P is the pressure (N m^{-2}), M is the mass (kg), T is the temperature (K), and k_{B} is the Boltzmann constant.

If we assume a unit sticking probability of one and a coverage of 10^{19} m^{-2}, the time taken for one monolayer of gas to cover the surface is:

$$\text{Time (s)} = \frac{10^{19}}{F}. \tag{2.10}$$

Using the molecular mass of air for M in equation (2.9), the fluxes, F, of gas molecules for increasing degrees of vacuum are shown in table 2.4 along with the corresponding time taken for a monolayer (ML) of adsorbate to form. At the very

Table 2.4. The amount of time taken for a monolayer (ML) of adsorbate (dry air) to form on a surface at varying vacuum pressures.

Degree of vacuum	Pressure (mbar)	Flux (m^{-2} s^{-1})	Time/ML (s)
Atmospheric	1000	2.8×10^{27}	3.5×10^{-9}
Low	1	2.8×10^{24}	3.5×10^{-6}
Medium	10^{-3}	2.8×10^{21}	3.5×10^{-3}
High	10^{-6}	2.8×10^{18}	3.5
Very high	10^{-9}	2.8×10^{18}	3.5×10^{3}
Ultrahigh	10^{-10}	2.8×10^{14}	3.5×10^{4}

high vacuums used in SIMS instrumentation, the time taken is almost 10 h, compared to only 3.5 s at 10^{-6} mbar, which is the typical vacuum pressure found in SEM instrumentation.

References

[1] Hellborg R and Whitlow H J 2019 *Electrostatic Accelerator: A Versatile Tool* (San Rafael, CA: Morgan and Claypool) 12–3

[2] Smith N S, Tesch P P, Martin N P and Kinion D E 2008 *Appl. Surf. Sci.* **255** 1606–9

[3] Smith N S, Notte J A and Steele A V 2014 *MRS Bull.* **39** 329–35

[4] Dowsett M G, Smith N S, Bridgeland R, Richards D, Lovejoy A C and Pedrick P 1997 *Secondary Ion Mass: Spectrometry SIMS X 10th International Conference on Secondary Ion Mass Spectrometry* ed A Benninghoven *et al* (Chichester: Wiley) p 367

[5] Cornett D S, Lee T D and Mahoney J F 1994 *Rapid Commun. Mass Spectrom.* **8** 996–1000

[6] Gillen G and Roberson S 1998 *Rapid Commun. Mass Spectrom.* **12** 1303–12

[7] Weibel D, Wong S, Lockyer N, Bleinkinsopp P, Hill R and Vickerman J C 2003 *Anal. Chem.* **77** 1754–64

[8] Ninomiya S, Ichiki K, Yamada H, Nakata Y, Seki Y, Aoki T and Matsuo J 2009 *Rapid Commun. Mass Spectrom.* **23** 3264–8

[9] Shao Y, Chen T C, Fenner D B, Moutsakas T D and Chu G 2003 *MRS Online Proc. Libr.* **743** 310

[10] Yamada I 2014 *Appl. Surf. Sci.* **310** 77–8

[11] Touboul D, Halgand F, Brunelle A, Kersting R, Tallarek E, Hagenhoof B and Laprevote O 2004 *Anal. Chem.* **76** 1550–9

[12] Kollmer F 2004 *Appl. Surf. Sci.* **231–2** 153

[13] Brunelle A, Touboul D and Laprevote O 2005 *Mass Spec.* **40** 985–99

[14] Liebl H 1980 *Scanning* **3** 79–89

[15] Magee C W, Harrington W L and Honig R E 1978 *Rev. Sci. Instrum.* **49** 477–85

[16] Grehl T 2003 Improvement in TOF-SIMS instrumentation for analytical application and fundamental research *Doctoral Thesis* Universität Münster, Münster

[17] Vanbellingen Q, Elie N, Eler M J, Della-Negra S, Touboul D and Brunelle A 2015 *Rapid comm. Mass Spectrom* **29** 1187–95

[18] Sigmund P 1969 Theory of sputtering. I *Phys. Rev.* **184** 383–416

[19] Matsunami N, Yamamura Y, Itikawa Y, Itoh N, Kazumata Y, Miyagawa S, Morita K, Shimizu R and Tawara H 1984 *Data Nucl. Data Tables.* **31** 1–80

[20] Seah M P 2005 *Nucl. Instrum. Methods Phys. Res.* B **229** 348–58

[21] Yamamura Y and Shindo S 1984 *J Radiat. Eff.* **80** 57–72

[22] Sigmund P and Claussen C 1981 *J. Appl. Phys.* **52** 990–3

[23] Jakas M, Bringa E and Johnson R 2002 *Phys. Rev.* B **65** 165425

[24] Seah M P 2008 *J. Vac. Sci. Technol.* A **26** 660–7

[25] Postawa Z, Paruch R, Rzeznik L and Garrison B J 2013 *Surf. Interface Anal.* **45** 35–8

[26] Seah M P 2013 *J. Phys. Chem.* C **117** 12622–32

[27] Cumpson P J, Portoles J F, Barlow A J and Sano N 2013 *J. Appl. Phys.* **114** 124313

[28] Seah M P, Havelund R and Gilmore I S 2016 *J. Am. Soc. Mass. Spectrom.* **27** 1411–8

[29] Seah M P, Spencer S J and Shard A G 2015 *J. Phys. Chem.* B **119** 3297–303

[30] Bertolini S and Delcorte A 2023 *Appl. Surf. Sci.* **631** 157487

[31] Lorenz M, Shard A G, Counsell J D P, Hutton S and Gilmore I S 2016 *J. Phys. Chem.* C **120** 25317–27

[32] Lee J L S, Gilmore I S, Seah M P and Fletcher I W 2011 *J. Am. Soc. Mass Spectrom.* **22** 1718–28

[33] De Hoffmann E, Charette J and Stroobant V 2007 *Mass Spectrometry: Principles and Applications* (Chichester: Wiley)

[34] Murray K 2022 *J. Am. Soc. Mass. Spectrom.* **33** 2342–7

[35] Irion M P, Selinger A and Wendel R 1990 *Int. J. Mass Spectrom. Ion Process.* **96** 27–47

[36] Smith D F, Robinson E W, Tolmachev A V, Heeren R M A and Pasa-Tolic L 2011 *Anal. Chem.* **83** 9552–6

[37] Makarov A 2000 *Chem.* **72** 1156–62

[38] Kingdon K 1923 *Phys. Rev.* **21** 408

[39] Passarelli M K *et al* 2017 *Nat. Methods* **14** 1175

[40] Zhang J T, Brown J, Scurr D J, Bullen A, MacLellan-Gibson K, Williams P, Alexander M R, Hardie K R, Gilmore I S and Rakowska P D 2020 *Anal. Chem.* **92** 9008–15

[41] Kotowska A M, Trindade G F, Mendes P M, Williams P M, Aylott J W, Shard A G, Alexander M R and Scurr D J 2020 *Nat. Commun.* **11** 5832

[42] Edney M K, He W, Smith E F, Wilmot E, Reid J, Barker J, Griffiths R L, Alexander M R, Snape C E and Scurr D J 2022 *Analyst* **147** 3854–66

[43] Vickerman J C 2001 SIMS, time-of-flight, and surface analysis *Encyclopedia of Materials: Science and Technology* ed K H Jürgen Buschow, R W Cahn, M C Flemings, B Ilschner, E J Kramer, S Mahajan and P Veyssière (Amsterdam: Elsevier) pp 8624–8

IOP Publishing

Secondary Ion Mass Spectrometry and Its Application to Materials Science (Second Edition)

Sarah Fearn

Chapter 3

Experimental ion beam considerations

Secondary ion mass spectrometry (SIMS) is an incredibly flexible analytical technique. Not only can chemical data be generated from the surface of a sample to provide mass spectra and ion maps, but depending on the selection of the primary ion beam conditions, material can also be precisely sputtered away to reveal compositional changes in the z dimension, providing a depth profile through the target material. Judicious selection of the ion beam species can also enhance the formation of particular secondary ion species or the collection of larger molecular species. Depending on the data required, ion beam conditions such as beam type, current, and energy need to be carefully considered, and the most appropriate selection of parameters chosen to provide the best data. In some cases, perhaps due to a lack of target material or just time, a compromise may well have to be made; therefore, beam optimisation is essential in such cases.

The formation of secondary ions can be described by the basic SIMS equation:

$$I_s^x = I_p \ C_x S_x \gamma F, \tag{3.1}$$

where I_s^x is the secondary ion count of species x, I_p is the primary ion beam current, C_x is the concentration of species x being measured, S is the sputter yield of species x, γ is the ionisation efficiency of the species being detected, and F is the instrument transmission factor.

As the ionisation of the sample occurs at or close to the emission of the sputtered particles, the chemistry of the sample also has an influence on the electronic processes involved. The sputter yield and ionisation efficiency are thus influenced by the chemical state of the sample. In addition to this, physical and chemical effects occur, depending on the type of primary ion beam used, and these influence secondary ion formation. In order to select the most appropriate ion beams for an experiment, it is important to understand these ion beam interactions. This next section will consider the parameters related to ion beam species,

doi:10.1088/978-0-7503-3331-3ch3

dose, and energy with regard to performing a SIMS experiment and the pros and cons that have to be considered.

3.1 Ion beam selection

3.1.1 Species

3.1.1.1 Atomic ion beams

As mentioned in section 2.1, ion beam sources can be produced using either gaseous or metal species. Depending on the technology used to form the ion beam, each type has its own physical characteristics, such as beam focus and brightness, for example. However, along with these physical specifications, an ion beam also has chemical attributes, which may or may not be useful in a SIMS experiment. It is therefore important to know what these are so that they may be either exploited or avoided, as the case may be. Reactive ion species such as oxygen and caesium have very specific effects upon a target material, and these can affect the ionisation efficiency and secondary ion yield of the element or molecule being sputtered by the ion beam.

The formation of a secondary ion is reliant on the ionisation efficiency γ, which is the ease or tendency of an element or molecule to form either a positive or negative secondary ion. Therefore, γ is dependent on the ionisation energy or electron affinity of the ion being measured. As one moves from the top to the bottom of the periodic table, elements in groups I and II have decreasing ionisation energies and more readily form positive ions: $X \Rightarrow X^+ + e$. However, this also varies hugely across the periodic table (see figure 3.1, which shows how the formation of positive ions varies by five orders of magnitude across the periodic table [1]). Conversely, elements that

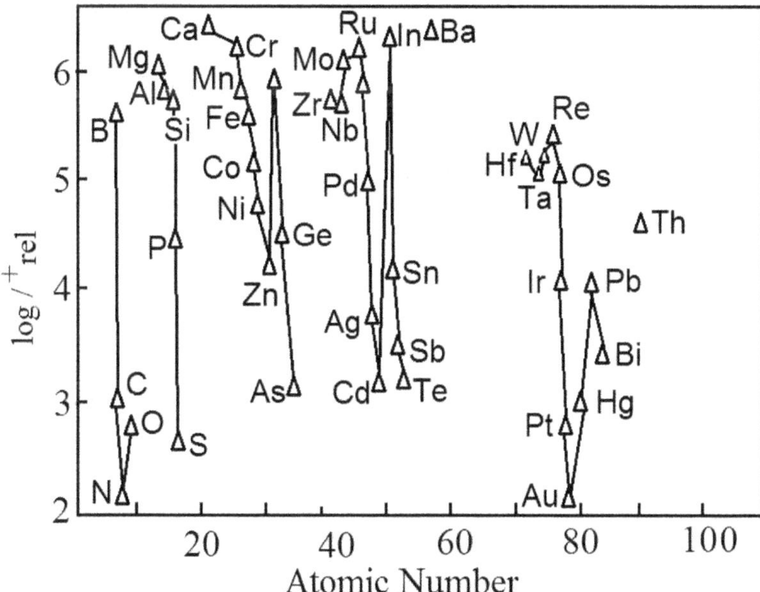

Figure 3.1. Variation of positive ion yield as a function of atomic number. Reprinted with permission from [1]. Copyright (1997) American Chemical Society.

have high electron affinities are more likely to form negative ions. Such elements are situated on the far right-hand side of the periodic table; for example, the halogens (F, Cl, Br, and I) readily form negative ions, with fluorine being the most reactive.

Together with the variation of elemental ionisation across the periodic table, the ion beam used to sputter the target material also has a chemical effect on the process.

The chemical changes brought about by using reactive primary ion beams such as oxygen and caesium can have some advantages. The use of an oxygen ion beam (which oxidises a sample) or the presence of oxygen in the form of an oxide layer enhances the formation of positive secondary ions in a SIMS analysis. Figure 3.2 shows how the secondary ion yields of metal ions can be increased by up to a factor of ~100 when detecting the same metallic species if oxygen is present [2]. Ion species that more readily form positive ions in the presence of oxygen therefore have their positive secondary ion yields increased further.

Typically, an increase in secondary ion yield is to be welcomed, as so few ions are formed, and in some cases, this increase in counts can help to improve the detection of some species. This enhancement of the secondary ion signal, however, may also be problematic. For example, lithium and sodium have no problem forming positive secondary ions; consequently, further enhancement of their secondary ion yield may saturate the detection system (depending on the concentration within the sample matrix), rendering the ion count statistically meaningless.

Conversely, for elements with a high electron affinity that more readily form negative secondary ions, it has been found that caesium primary ion beams enhance secondary ion yields because caesium lowers their work function. Again, as the ease

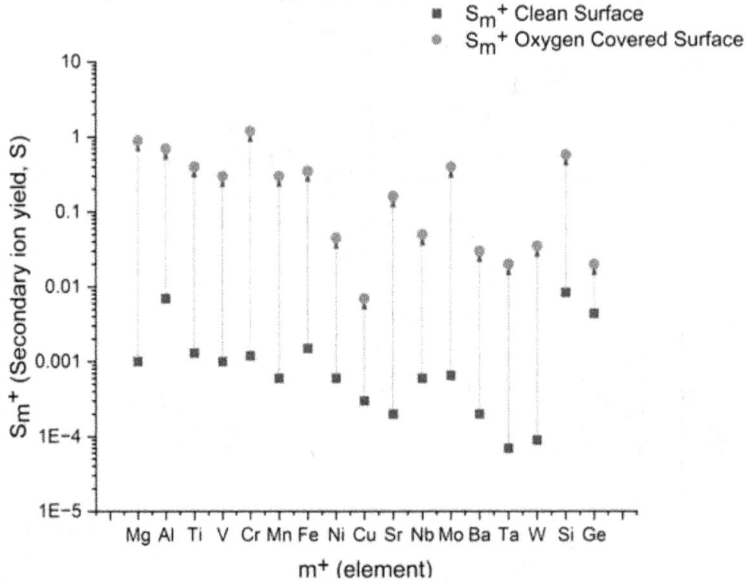

Figure 3.2. Increase in the detected positive secondary ion yield for metal ion species produced by clean surfaces compared to those detected in the presence of oxygen [2].

of ionisation and the formation of negative secondary ions vary from species to species, this enhancement is not always beneficial. For example, species such as fluorine and oxygen readily form negative secondary ions. Under caesium ion beam bombardment, ion yield enhancement again leads to the saturation of the detection system. In this case, other ions such as $^{32}O_2^-$ may be more suitable ions to collect.

As a general rule, positive secondary ions are promoted by the presence of a strongly electronegative element such as O, either implanted or adsorbed on the sample surface; conversely, a strong electropositive element such as Cs promotes negative secondary ion emission, as highlighted in figure 3.3. The variability in the sputter yields S and ionisation efficiency γ due to the ion-beam-induced chemical changes is known as the **matrix effect**.

Another important consideration when using reactive ion beams such as oxygen or caesium is the phenomenon known as ion-beam-induced segregation. This arises from the chemical effects of the ion beam mixing with the target matrix and forming an altered layer in the sample and the specific secondary ion being collected. The phenomenon is identified by SIMS profiles with either grossly extended or shortened depth profiles and is discussed and explained in more detail in section 3.2.3.

It is entirely possible to avoid the chemical effects described above by using a primary ion beam formed from an inert species such as argon (Ar) or xenon (Xe). In this case, ion beam mixing of the primary ion beam and target still occurs, but the inert primary ion beams do not enhance either positive or negative secondary ion formation. When using inert primary beams, it is still important to recognise the tendencies that elements have to form ions and the wide-ranging sensitivity that occurs between elements in SIMS. A good illustration of this is an analysis of

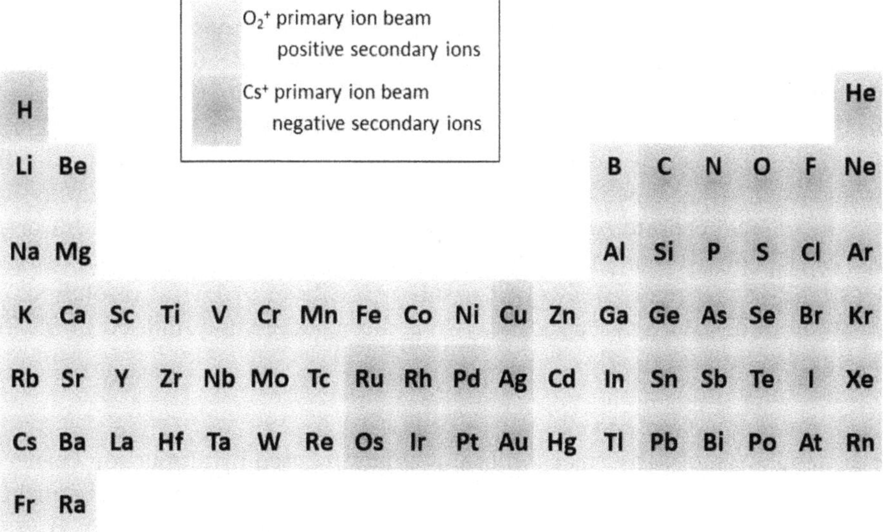

Figure 3.3. General rule for primary ion beam selection for the detection of elemental species across the periodic table.

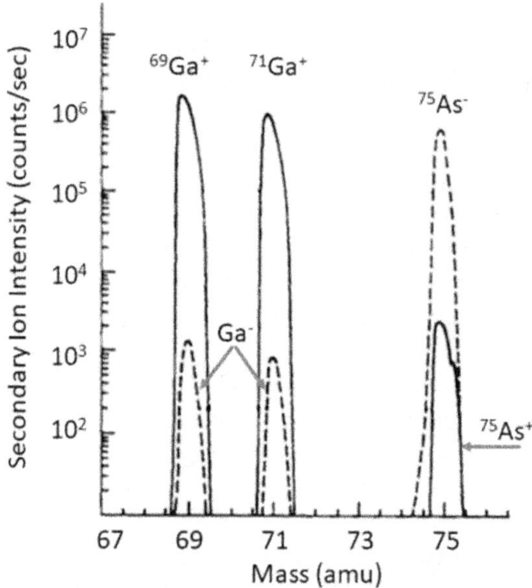

Figure 3.4. Positive and negative SIMS spectra from GaAs obtained using an Ar^+ primary ion beam, highlighting the differing sensitivities to elements in the same matrix. Reprinted with permission from [3], Copyright (1981), with permission from Elsevier.

gallium arsenide (GaAs) using an argon ion beam (Ar^+) (see figure 3.4). Without understanding that elements ionise differently across the periodic table—and that gallium and arsenic may not ionise equally—measuring only positive secondary ions could give misleading results. Specifically, the data obtained might suggest that the sample contains mainly Ga, with only a trace amount of As present. The opposite would also have been the case if only negative secondary ions had been collected. This result nicely highlights the different sensitivities in SIMS for different elements in the same matrix [3].

In the case of detecting molecular secondary ions, liquid metal ion gun (LMIG) sources are now widely used, particularly in time-of-flight (ToF)-based SIMS instruments. LMIGs based on bismuth (Bi) and its small clusters, such as $Bi_3{}^+$, $Bi_3{}^{2+}$, $Bi_5{}^+$, etc. generate better secondary ion yields compared to Ga, In, or Au-based LMIGs [4, 5]. Using the very small probe size of the LMIG, it is possible to generate SIMS ion maps well into the submicron level of lateral resolution. The application of LMIG focused ion beams has been successfully utilised in SIMS to microanalyse surfaces [6, 7].

3.1.1.2 Polyatomic ion beams

The accumulation of subsurface ion beam damage caused by atomic ion beams has limited SIMS to the surface layer in molecular analysis. The development of polyatomic sources such as Ar gas cluster ion beams (GCIBs) and H_2O has broadened the range of materials that can be successfully analysed in SIMS. Polyatomic clusters have reduced impact energies, resulting in reduced subsurface damage and increased molecular ion yields. These sources are now routinely used for the depth profiling of polymer electronics as well as biological analysis [8–10]. Argon-based GCIBs have relatively good secondary ion yields; however, their ability to clearly resolve peaks and their mass accuracy have been limitations [11]. To maintain the integrity of molecular samples as much as possible under ion beam bombardment, GCIBs have become the preferred ion beam species for organic analyses. More details of the ion beam interactions taking place can be found in section 3.2.

3.1.2 Energy

After selecting the ion beam species to be used in a SIMS experiment, the ion beam energy must be selected. Typically, the energy of the ion beam varies from the low-energy range of ~0.2 keV up to 2 keV, or it may be in the higher range of 5–30 keV. The specific energy of the ion beam depends on the type of SIMS experiment to be carried out and the features to be resolved; for example, a deeply buried interface, a surface coating, or the distribution of a particular secondary ion over the x–y plane, or all three dimensions.

Typically, the higher the ion beam energy, the greater the beam focus and beam currents produced. Reduced ion beam energies, however, have the important benefit of inducing much lower ion beam mixing of the target material. These ion beam interactions are discussed in more detail in section 3.2, but some general features associated with ion beam energy are presented here.

During the ion beam bombardment of the sample, the primary ion beam becomes implanted into the target material, mixes up the matrix of the sample, and forms what is known as the altered layer (see the schematic in figure 3.5). The thickness of the altered layer, x, is dependent on the projected range of the primary ions, R_p, and the angle of incidence, θ, of the primary ion to the surface normal, such that:

$$x = R_p \cos\theta. \tag{3.2}$$

Using the free software Sputtering Range of Ions in Materials (SRIM) [12], a table of projected ion ranges R_p for different ion beam species and different energies impacting a known target material, such as silicon, at normal incidence can be created (see table 3.1). Such a table highlights the reduction in R_p associated with reducing the ion beam energy. With the calculated R_p values from SRIM and using equation (3.2), the altered layer x formed at any angle to the surface normal can then be estimated.

SRIM allows the simulation of a SIMS analysis to be carried out. For example, a layered structure of Si/SiGe/Si is to be analysed to identify how sharp the interfaces between the layers are. Each layer in the structure is 10 nm thick. The beam conditions proposed are a 2 keV oxygen ion beam at normal incidence ($\theta = 0°$) to the sample surface. The ion range for these conditions is shown in figure 3.6(a). Under these ion beam conditions, it can be seen that the ion beam mixing would extend beyond the thickness of the top layer, and the interfaces would not be resolved. If we were to reduce the ion beam energy to 1 keV and use an incidence angle of 45° to the

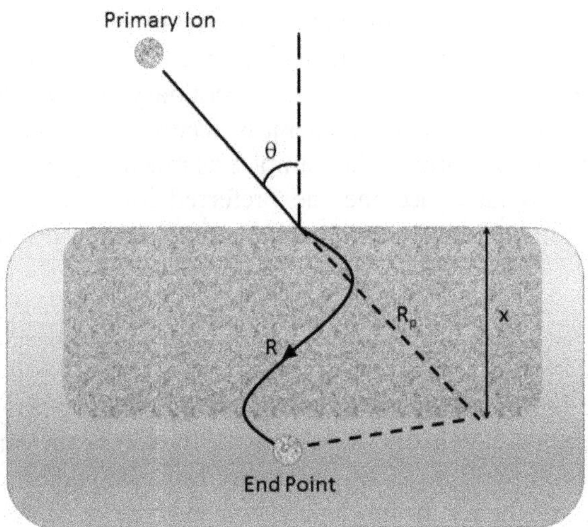

Figure 3.5. A schematic showing the thickness of the altered layer x formed in a sample under ion bombardment. Also shown is R, the path length of the primary ion, and Rp, the projection of the primary ion following the path of initial impact at an angle of θ to the surface normal. The primary ion comes to rest at the end point.

Table 3.1. Estimation of the projected ion ranges R_p of O^+, Ar^+, and Cs^+ ions into silicon at normal incidence for different ion beam energies.

	Projected ion range, R_p (nm)		
Ion energy (keV)	Beam: O_2^+	Beam: Ar^+	Beam: Cs^+
0.5	1.8	2.3	3.0
1	2.7	3.4	4.0
1.5	3.5	4.2	4.8
2	4.3	5.0	5.5
2.5	5.0	5.8	6.0
3	5.8	6.5	6.6
4	7.1	7.8	7.6
5.0	8.5	9.1	8.4

surface normal, the SRIM model (figure 3.6(b)) shows that the ion range would be contained within the top layer of the sample and the top interface would just be resolved. It should be noted that SRIM only simulates the physical effects of the ion beam bombardment and not any of the chemical effects of the ion beam on the target materials. Along with the collision plots, many other parameters are also estimated in the programme, such as energy loss to phonons, energy loss to recoils, and target vacancies, to name a few.

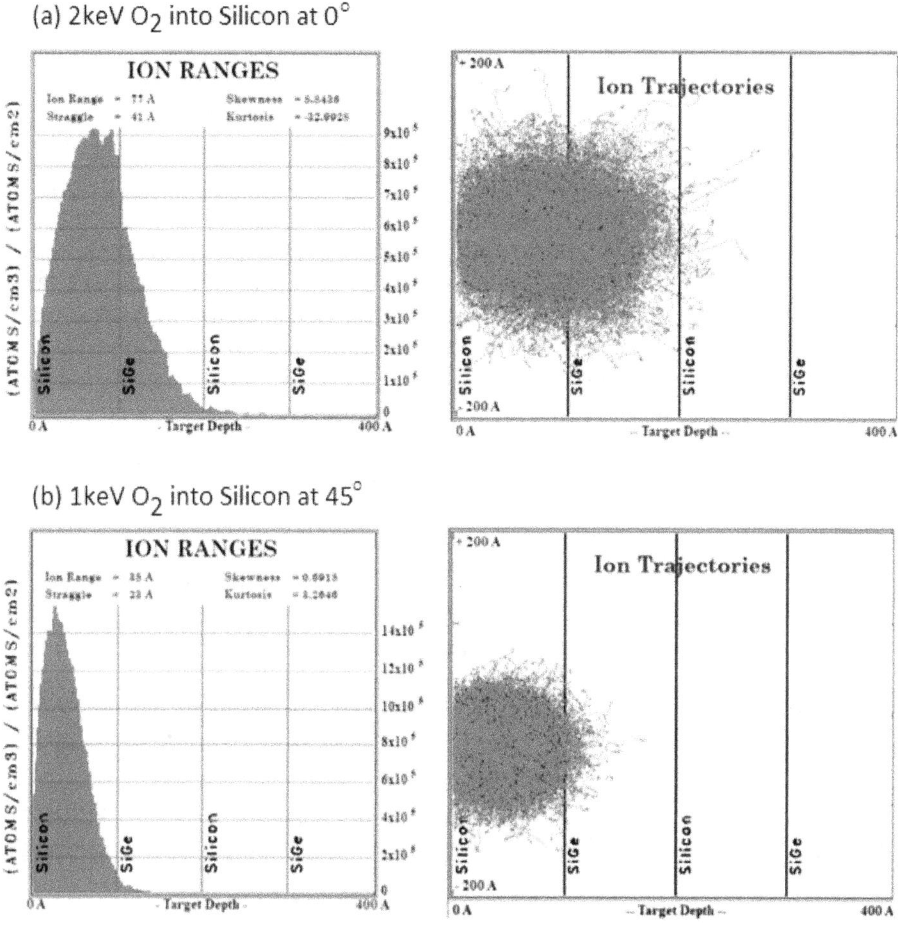

Figure 3.6. SRIM simulations of an oxygen ion beam bombarding a Si/SiGe/Si/SiGe layered structure at (a) 2 keV and 0° incidence and (b) 1 keV and 45° incidence.

3.1.3 Dose

The ion beam dose is dependent on the current of the ion beam and the duration for which the ion beam irradiates the selected region of the target. The ion beam dose used during a SIMS analysis defines whether the analysis is either *static* or *dynamic* SIMS.

In static SIMS, a very low primary ion beam current is used, and no part of the analytical area is impacted more than once by the ion beam. Under these conditions, static SIMS analysis is considered to be non-destructive. To remain in this static regime, a limited dose must be applied to the sample; this is known as the *static limit.* For a well-defined crystalline material, this limit is an ion beam dose of approximately 10^{15} ions cm^{-2}. The static limit varies from material to material, and for many polymers, molecular, and biological samples, the static limit is unknown. As their structures are more open and have lower binding energies, their static limits are much lower than those of crystalline materials; they can be as low as 10^8 ions cm^{-2} and not greater than 10^{12} ions cm^{-2} [13].

Once the ion beam dose goes beyond the static limit, the analysis is then considered to be dynamic SIMS. At this point, as the ion beam rasters over the analytical region, material is sputtered away. After a period of time, a sputter crater forms that is visible on the sample. The rate at which the crater forms depends on the ion beam current and the size of the sputter crater. The sputter rate increases linearly with current, whereas its dependence on the sputter area is proportional to $1/(\text{area})^2$. Thus, halving the sputter area would lead to a fourfold increase in sputter rate.

As the ion beam dose is critical for understanding the type of SIMS experiment being performed, an accurate measurement of the beam current is essential at the start of the SIMS analysis, as this is an essential parameter for calculating the ion beam dose. It is also important to measure the ion beam current at the end of an analysis to ensure that the beam has been stable over the analytical time.

3.1.3.1 Calculating ion beam dose

The ion beam parameters needed to calculate the ion beam dose are the ion beam current, sputter crater size, and the dwell time of the ion beam on the sample. Calculating the dwell time depends on whether the ion beam is continuous or pulsed as it irradiates the sample. Typically, quadrupole and magnetic sector analyser SIMS instruments use continuous beams, while time-of-flight-based SIMS instruments have pulsing ion beams.

For example, a continuous 1 keV Ar^+ ion beam with a measured current of 60 nA is rastered over an area of 100 μm^2. The time of the analysis is 20 min. The number of charged particles supplied by the ion beam to the sample surface during the experiment is therefore:

$$\text{no. of charged particles} = \frac{\text{current} \times \text{time}}{\text{charge on electron}} = \frac{i \times t}{e}. \tag{3.3}$$

Under the conditions described above:

$$\text{no. of charged particles} = \frac{60 \times 10^{-9} \times 1200}{1.6 \times 10^{-19}} = 4.5 \times 10^{14}.$$

The ion beam dose, therefore, is the number of charged particles supplied during the experiment divided by the analytical area exposed to the ion beam. Under the conditions outlined above, the ion beam dose is:

$$\text{ion beam dose} = \frac{\text{number of charged particles}}{\text{area}} \tag{3.4}$$

$$\text{ion beam dose} = \frac{4.5 \times 10^{14}}{0.01 \times 0.01} = 4.5 \times 10^{18}\ \text{ions cm}^{-2}.$$

The ion beam dose calculated in the above example clearly describes a dynamic SIMS experiment, as it is above the static limit of 10^{15} ions cm^{-2}, and material would therefore be removed/sputtered from the target.

When the ion beam is pulsed over the sample surface, as in a time-of-flight-based SIMS instrument, the dwell time of the ion beam on the sample is a little more complicated to calculate. For example, the ion beam is rastered over a selected area of 100 μm^2 at a measured current of 1 pA. As the beam pulses over the area, there are 128 × 128 pulses in each scan, and this is repeated 50 times (50 scans). To calculate how long the ion beam dwells on the sample over the whole length of the SIMS analysis, it is necessary to know the length of the individual pulse (this parameter is often called the beam width). For this example, the beam width is set to 14 ns. The dwell time of the ion beam for each scan is therefore:

$$\begin{aligned} \text{dwell time for each scan} &= (\text{no. of pulses}) \times (\text{beam width}) \\ &= (128 \times 28) \times (14 \times 10^{-9}) = 2.29 \times 10^{-5}\,\text{s}. \end{aligned}$$

This is repeated for 50 scans, so the total beam dwell time during the analysis is 1.15×10^{-3} s. Now that the time is known, the ion beam dose can be calculated using the same equations as shown in the previous example. The result gives a dose of 7.17×10^7 ions cm^{-1}, which is well below the static limit and would remove very little material from the sample surface.

3.2 Ion beam interactions

The bombardment of a target with an ion beam causes both the removal and ionisation of the target atoms, along with a number of deleterious target–beam interactions shown schematically in figure 3.7 [14]. The deleterious ion beam–target interactions can distort the analyses obtained during a SIMS experiment. These interactions relate primarily to SIMS experiments where the ion beam dose exceeds the static limit and dynamic SIMS is being carried out (i.e. depth profiling). It is

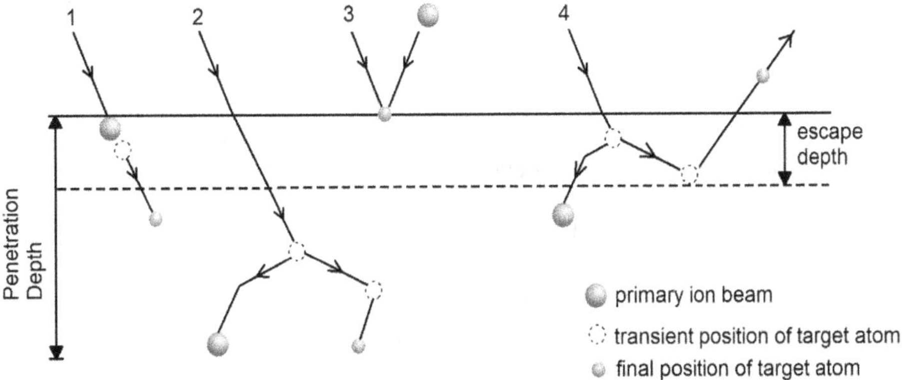

Figure 3.7. Schematic of the beam–target interactions occurring during SIMS analysis: 1. recoil implantation; 2. cascade mixing; 3. primary ion reflection; 4. cascade mixing with sputtering. The escape depth of the sputtered atoms is about two to three atomic layers in crystalline materials.

important to be aware of these deleterious ion beam processes and how they present themselves in the resulting data so that the correct and meaningful interpretation of the data is made.

The two major interactions between the ion beam and the target atoms are recoil implantation and cascade mixing. The ion beam hits the target surface and penetrates the near surface. The bombarding ions possess energy and momentum, which they share with the target atoms. Recoil implantation (number 1, figure 3.7) occurs through direct impact collisions between the ions and the target atoms. Only enough energy is given to the target atoms to knock them anisotropically into the solid, where they come to rest [15]. Cascade mixing (number 2, figure 3.7) occurs as the interactions between the primary ions and the resident atoms cause further collisions, creating an isotropic collision cascade. The cascade continues to develop until the transferable energies become less than the displacement energy of the target atoms [16]. In both cases, any excess energy is released as phonons. The primary beam species comes to rest at various depths within the host lattice. The penetration depth of these ions is dependent on the beam energy and the angle of incidence of the primary beam with respect to the target surface.

An atom or group of atoms may receive enough energy in a suitable direction to enable it to overcome the surface binding forces and be sputtered from the target (number 4, figure 3.7). The approximate depth of origin of these emitted atoms and ions is about two to three atomic layers for inorganic crystalline materials [17]; however, as mentioned earlier, the location of the emitted particles can be up to 10 nm away from the initial impact of the primary ion [18]. The potentially small escape depth of the secondary ion signal is further degraded due to the deleterious ion beam–target interactions. Some of these effects will be discussed briefly in the following section, but more detailed information can be found in the literature [2, 19].

3.2.1 Ion-beam-induced atomic mixing

Cascade mixing and recoil implantation are the most well-known factors limiting depth resolution (see below for more details), giving rise to the broadening of buried features and interfaces. Out of the two processes, it has been suggested that cascade mixing is the dominant material transport process accompanying sputtering [20–22].

The collision cascade initiated by the primary ion impact is capable of efficiently mixing surface and subsurface layers over a distance related to the order of the primary ion range, R [23]. The efficiency of the cascade mixing is such that the mixed zone can propagate through the material faster than the sputtering erodes the surface. At this point in a SIMS analysis, the ion beam is essentially now analysing a mixed layer composed of the original host material and the implanted ion beam species [8]. The mixed layer (or altered layer, as it is also known) is continuously pushed into the substrate ahead of the sputtering front, resulting in final depth profiles that decay exponentially, leading to asymmetric depth profiles [24]. The influence of cascade mixing on SIMS depth profiles can, however, be reduced by lowering the energy of the primary beam, using heavy ions, employing glancing angles, or combining these approaches.

3.2.1.1 Depth resolution

The degree of cascade and recoil mixing evidenced in SIMS analyses is assessed using the depth resolution parameter. For this parameter to be both transferable between analyses and meaningful, it needs to be defined. If we imagine performing a depth profile on a very abrupt implant within a sample, such as a delta layer (which is composed of an atomic layer of atoms in a host matrix; see figure 3.8(a)), we can envision what the ideal profile would look like if there were no ion beam mixing events. In this case, the ideal SIMS profile would follow the red dotted line shown in figure 3.8(b). The interface would be abrupt on either side of the delta layer, rising instantaneously as the implant layer was reached and dropping immediately to zero once the layer was passed. However, in reality, the depth profile resembles the black line shown in figure 3.9. The profile is an asymmetric Gaussian shape, with a sharper up-slope known as the leading edge, λ_l, on the approach to the implant, and a broader down-slope known as the trailing edge, λ_t, as the beam sputters through the implant feature.

The decay of the trailing edge is commonly used to assess implant profiles and layered structures. Past the position of the apparent peak of the profile, Z_o, the trailing edge typically exhibits an exponential decay. The depth resolution, λ_t, of this slope is thus measured as the depth, ΔZ, across which the trailing edge drops by a specified amount [15]. Many ways of measuring λ_t have been used, and some are defined below:

1. The depth over which the signal drops by a decade (nm/decade). See figure 3.8(b).
2. The depth over which the signal drops by a factor of ten; e.g. from $0.1C_m$ to $0.01C_m$.
3. The depth over which the signal drops to $1/e$ of its original value.

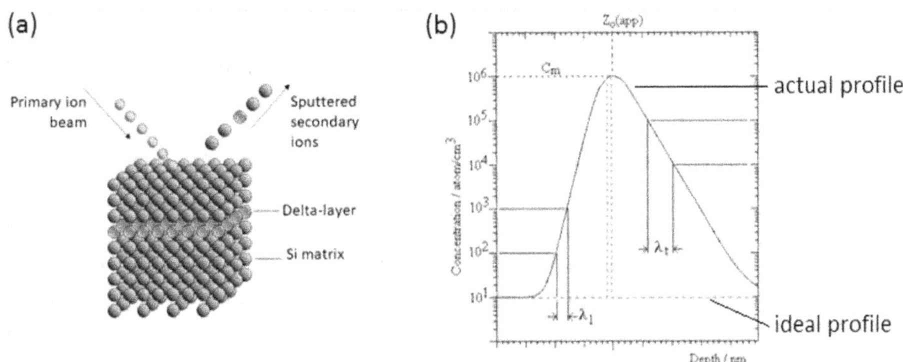

Figure 3.8. Schematic of (a) a delta-layer structure and (b) a SIMS profile of this feature, highlighting the leading and trailing edges, λ_l and λ_t respectively, on the SIMS profile. This illustrates the practical depth resolution compared to the ideal profile (red dotted line) without beam mixing. The apparent peak position Z_o is also shown.

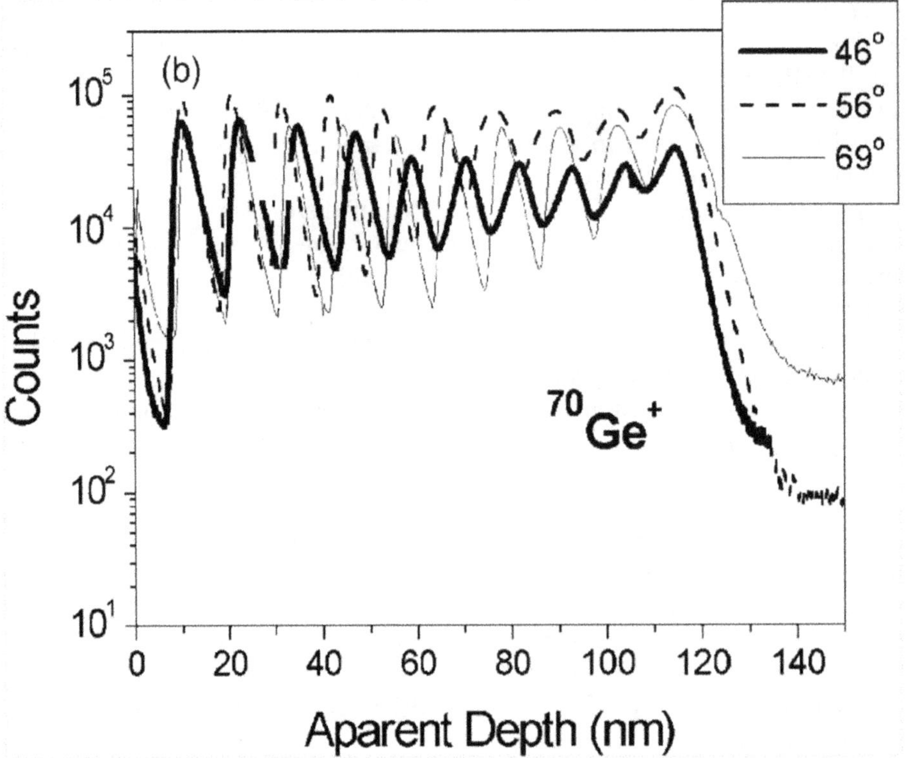

Figure 3.9. SIMS depth profile through a SiGe δ-doped sample. The onset of roughening and loss of depth resolution occurs at different depths for the different angles of ion beam incidence on the sample. Reprinted with permission from [26], Copyright (2003), with permission from Elsevier.

When analysing an abrupt interface, the value of λ_t between 84% and 16% of the maximum value is also a very popular measure of depth resolution, as it is directly related to the standard deviation of the corresponding Gaussian distribution, so that $\Delta Z = 2\sigma$ [25].

The up-slope, or leading edge, is also a useful measurement when used in combination with λ_t. It is measured in the same way as λ_t, has the same units, and is denoted by λ_l. The experimental values obtained for λ_t are often greater than the corresponding λ_l values, indicating the predominance of ion beam mixing in the sample along the direction of ion beam impact, as opposed to upwards. This results in the asymmetric nature of the profiles obtained. Depth resolutions are typically between 3 and 20 nm/decade on the trailing edge and possibly more, depending on the ion beam energy and initial surface topography.

The width of the peak at half its peak concentration value can also be measured. This is the full width at half maximum (FWHM). This measurement, however, does not provide any information about the shape of the profile and assumes a symmetric profile, which is rarely the case.

3.2.2 Beam-induced surface roughening and uneven etching

Another major influence on SIMS analyses is the effect of surface roughening and uneven sputtering. Topographical evolution and uneven etching are observed in the region sputtered by the ion beam, which results from nonuniform sputtering. The effects of roughening are particularly evident in multilayer structures. The depth resolution typically degrades with depth, and the angle of incidence of the ion beam can also have an influence, as highlighted in figure 3.9, which shows a depth profile through a SiGe δ-doped sample. As the sample surface roughens, the SIMS depth profile through a multilayered structure displays increasingly broad peaks [26]. The peak-to-valley dynamic range also degrades.

Induced topography, such as cone formation, has been known for some time [27, 28]. This phenomenon has been extensively investigated and found to be ion beam and sample dependent; that is to say, the topographical roughness of a large-grained metal differs from that of an amorphous material [29].

Topography can also develop due to preferential sputtering. In this case, one element is sputtered away more rapidly than another. The extent of preferential sputtering is influenced by several factors, including crystallisation and chemical properties. Preferential sputtering occurs along edge features and grain boundaries, leading to the formation of cones in the sputtered region [15]

In the case of uneven etching, adjacent layers may be exposed to the ion beam at the same time, producing an unresolvable depth profile. Using various SIMS instruments, the influence of uneven etching was empirically and theoretically studied [30]. It was found that as little as 0.1% unevenness of the crater base could influence a depth profile. A periodically doped multilayer structure was modelled with 1% and 10% unevenness on the crater base. The results showed that the inter-peak dynamic range decreased with depth and was bounded by amplitude envelopes. In the case of the crater with 1% unevenness, the envelopes met after 100 peaks, whereas for the crater with 10% unevenness, they met at the end of ten peaks, with the 10th and 11th layers being unresolvable.

3.2.3 Beam-induced segregation

The use of certain reactive ion beams for SIMS analyses can improve the ionisation efficiency of the target material and therefore improve the analysis. During SIMS depth profiling with reactive ion beams such as oxygen, however, beam-induced segregation can occur. This phenomenon is identified by SIMS profiles with grossly extended decay lengths.

Various suggestions have been put forward to explain why this segregation occurs, but explanations based on thermodynamics and field-induced segregation have emerged as the most consistent mechanisms to explain the process.

The thermodynamic argument, or Gibbsian segregation [31], depends on the relative heat of formation of the compound formed during reactive sputter profiling. This process can be most easily explained by considering copper-implanted silicon under oxygen bombardment at normal incidence. Upon sputtering, an altered surface layer of SiO_2 is initially formed as the beam reacts with the sample. The relative heat of formation of SiO_2 is -217 kcal mol^{-1}, whereas for CuO and CuO_2, the relative heats of formation are -37 and -40 kcal mol^{-1}, respectively [32]. Under oxygen bombardment, therefore, it is more energetically favourable for SiO_2 to be formed, and the Cu segregates away from the beam-induced oxide. The Cu is not sputtered efficiently, and the decay length for the impurity becomes grossly extended. Elements such as Ag, Pb, and Au were also found to have a similar segregation tendency in silicon [33]. Investigating beam-induced segregation effects caused by oxygen and nitrogen primary beam bombardment, Homma *et al* [34] found that in some cases, the reverse process 'improved' decay lengths. In situations where it was more energetically favourable for the impurity to oxidise rather than the silicon, the impurity segregated towards the beam-induced oxide and was more efficiently sputtered, leading to shorter decay lengths.

The field-induced segregation model also involves the altered layer that is formed during reactive sputter profiling. Using silicon under oxygen bombardment again as an example, the SiO_2 altered layer that is formed on the sample surface is insulating. Continued ion bombardment causes the surface to become charged, and an electric field is induced across the insulating layer. The effect of the induced field on impurity segregation was observed in the differing decay lengths obtained for Cu implants in *n*- and *p*-type silicon [35, 36]. The Cu was found to segregate more in the *n*-type silicon than in the *p*-type material. This was explained by the fact that the charge at the surface of the altered layer differs in strength for the two types of silicon; due to the presence of a depletion layer in the *p*-type silicon, the surface charge was found to be higher [37]. Thus, as the field across the altered layer decreases, the segregation effects increase. This process was highlighted further when analyses performed under electron bombardment to eradicate surface charging produced even longer decay lengths. Upon performing analyses with an O^- primary beam, the segregation effects were found to diminish [23]. Increased segregation effects under electron bombardment were also observed on As-implanted Si [38].

3.2.4 Polyatomic ion beam ion interactions

Polyatomic ion beam sources based on carbon fullerenes (C_{60}) have demonstrated potential for characterising biomolecular structures [39, 40]. These electron impact (EI) sources work by using charged particles comprised of multiple atoms, commonly known as a 'cluster ion beam'. The main advantage of a polyatomic ion source is that, upon impact with the target surface, as the large cluster breaks up, the kinetic energy of the polyatomic ion beam is distributed over a large number of atoms, and multiple localised low-energy impacts occur. The impact energy per ion is, therefore, reduced as opposed to the high-energy impacts that occur with atomic (e.g. Cs^+) or diatomic (e.g. $O_2{}^+$) ion beams [41–44]. The simulation of a $C_{60}{}^+$ ion beam impact on an amorphous ice layer on a silver substrate was compared to the ion impact of a $Au_3{}^+$ ion beam [45]. The results showed that a much wider crater is formed with the $C_{60}{}^+$ ion beam impact, and more material is ejected from the top layer of the target, coupled with a reduced damage cascade (i.e. mixing) in the underlying Ag substrate.

The result, therefore, of these multiple low-energy impacts is that ion beam damage to molecular structures is greatly reduced, and larger molecular ions and fragments can be collected with increased secondary ion yields because subsurface damage is reduced.

GCIBs based on Ar have now become very commonly used for the depth profiling of molecule-based materials [46–49]. The very large clusters of argon used in the ion beams are formed by the supersonic expansion of high-pressure gaseous argon through a nozzle, forming Ar_n^+ clusters ranging in size from $500 \leqslant n \leqslant 2500$ [50]. A simulation of a C_{60}^+ ion beam impact compared to the impacts of Ar_{18}^+, Ar_{1000}^+, and Ar_{2500}^+ clusters on a fullerene-based material is highlighted in figure 3.10 [51]. The simulation shows that the size and shape of the damaged areas caused by the C_{60}^+ and smaller Ar_n^+ cluster ions (when $n=18$) are similar, and the energy due to the ion impacts is deposited in the very near surface. As the argon cluster size is increased, the shape of the damaged region changes; the lateral size of the damage increases, whilst its depth in the target decreases. This significant change in the way in which the projectile interacts with the target material leads to both an increase in the ejection of intact molecules from the sputtered region and reduced fragmentation of the sputtered molecules. The use of these large cluster ion beams does not, however, lead to an increase in secondary ion yields [52], but as much less fragmentation occurs and more of the parent molecules are captured intact, a cleaner mass spectrum is produced [53]. This is very important for the differentiation of materials where the chemical bonding is the only difference in the sample, that is to say C–H bonded materials, as opposed to the detection of elemental changes in composition.

As Ar_n^+ GCIBs have been applied more and more in SIMS, understanding the underlying beam interactions has been imperative. An inter-lab study was carried out to investigate the depth profiling of an organic multilayer reference material using polyatomic ion beams [49]. An organic multilayer reference material made of

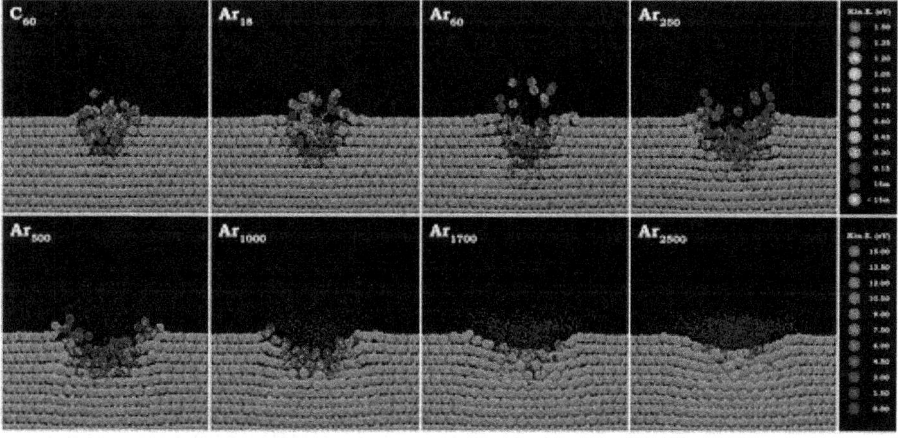

Figure 3.10. Simulations of C_{60}^+ and Ar_n^+ $(18 \langle n \rangle 2500)$ ion beam impacts on a fullerene substrate after 3 ps. Reprinted with permission from [51]. Copyright (2013) American Chemical Society.

an Irganox 1010 matrix with 1 nm marker layers of Irganox 3114 placed at depths of 50, 100, 200, and 300 nm was depth profiled. A schematic of the structure is shown in figure 3.11(a). The Ar_n^+ GCIB produced a depth profile with an improved and constant depth resolution compared to that of the C_{60}^+ ion source (see figure 3.11 (b)). The depth profiling of organic materials with Ar_n^+ cluster ion beams appears to have now superseded most depth profiling applications of the C_{60}^+ ion beam.

As with atomic and diatomic ion beams, the sputtering yield of polyatomic argon ion beams is an important parameter to understand. A previous study has shown that the sputter yield, Y, for an Ar_n^+ GCIB is linearly dependent on E, the energy of the ion beam [54]. In contrast, Y was found to decrease for increasing values of n, the number of argon atoms in the cluster, for a 20 keV Ar_n^+ ion beam [55]. Using data available in the literature, Seah was able to define a universal equation for argon gas cluster sputtering yields for eight materials: Au, SiO_2, Si, Irganox 1010, HTM-1 (a model organic light-emitting diode material), polystyrene, polycarbonate, and poly (methyl methacrylate) (PMMA) [56]. The sputter yield, Y, of atoms sputtered per primary ion is given by the universal equation:

$$\frac{Y}{n} = \frac{(E/An)^q}{[1 + (E/An)^{q-1}]}, \tag{3.5}$$

where the parameters A and q are obtained by fitting.

The combined data for all the materials studied is plotted in figure 3.12. The plot shows two important points: first, that the sputter rate for the organic materials is two to four times higher than for the elements or compounds such as SiO_2. Second, over the range of energies per argon atom (E/n) investigated, the plot shows that the sputter rate of both the organic and inorganic samples can change by a factor of 100. It was postulated, however, that for polymers with a high level of crosslinking, the sputter yield should decrease, as stronger bonds need to be broken.

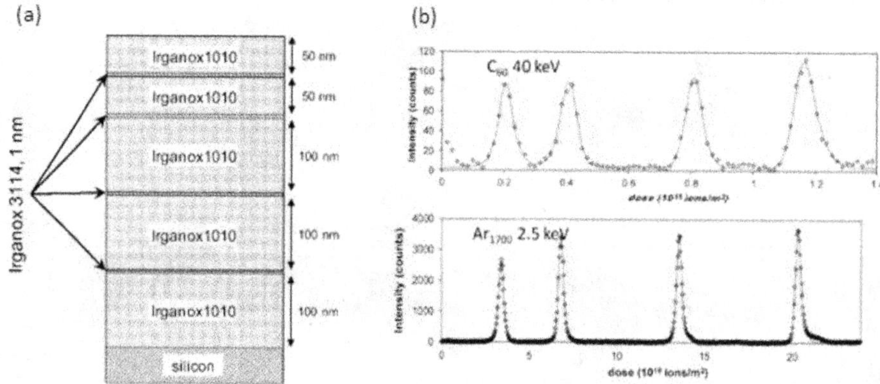

Figure 3.11. (a) Schematic of the structure of the multilayered sample used in the inter-lab study. (b) Depth profiles obtained using C_{60}^+ and Ar_n^+ cluster ion sources, plotting the $[M3114 - R]^-$ secondary ion intensities against the sputtering ion dose. Reprinted with permission from [49]. Copyright (2012) American Chemical Society.

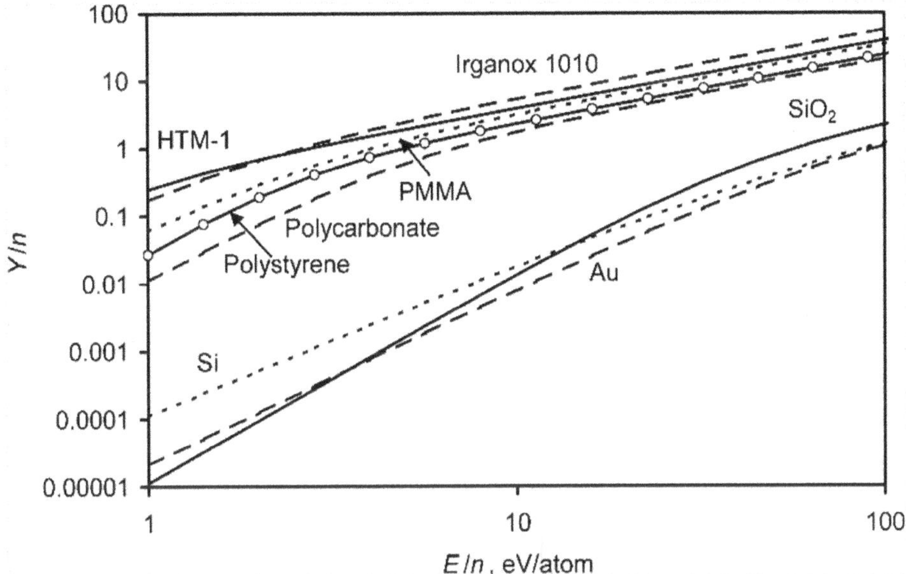

Figure 3.12. Compiled plot of *Y/n* for the eight materials analysed. The data was obtained at an incident angle of 45° except for PMMA and Si, where the data was obtained at an incident angle of 0°. Reprinted with permission from [56]. Copyright (2013) American Chemical Society.

3.3 Modes of SIMS analysis

3.3.1 Static mode

3.3.1.1 Mass spectra

High-resolution mass spectra are obtained from sample surfaces using the primary ion beam at an extremely low ion beam dose in static mode. The highest-resolution mass spectra are achieved with an ion beam pulsing on a predefined spot of the target surface with very short timescales. The pulsing is of the order of nanoseconds (~ns). The primary ion gun is then blanked off, and the generated secondary ions are extracted and accelerated into the time-of-flight analyser (figure 3.13). The secondary ions are then separated according to their mass-to-charge ratio (m/z) within the time-of-flight tube. The mass spectrum is recorded along with the ion beam spot coordinates where the secondary ions were generated. The primary ion beam is then moved to an adjacent pixel and the spot is irradiated by the short ion beam pulse. This process is repeated until the desired area (x, y) has been scanned and analysed, producing a mass spectrum. Due to the very short primary ion beam pulses, the mass resolution of the spectra is very high; typically, a resolving power, R, of 10 000 is routinely obtained.

Mass spectra from ToF mass analysers contain many peaks; it is therefore helpful to have some knowledge of the chemistry of the sample to at least, in the first instance, identify the main peaks of the sample, whether that be an element in an inorganic material or the parent molecule of an organic sample. A common strategy for accurate peak assignment is to make use of the stable isotopes of the species

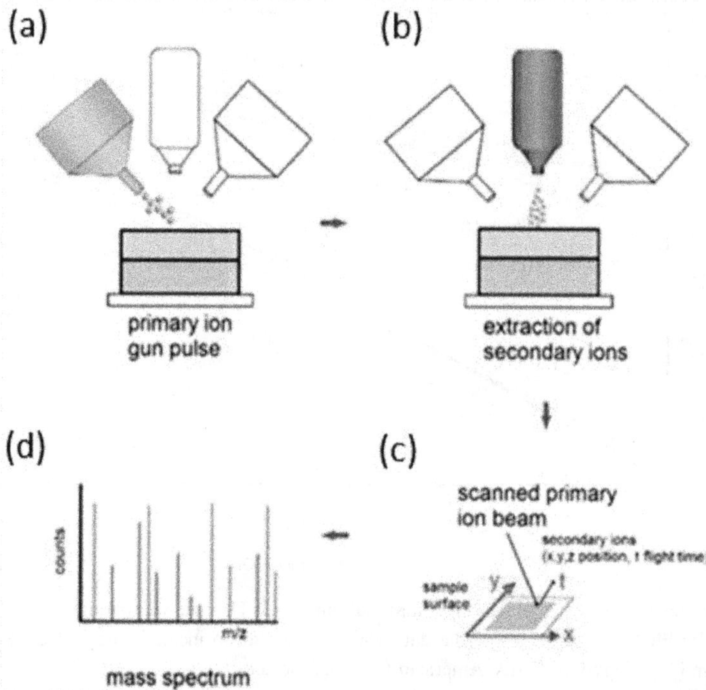

Figure 3.13. Steps needed to obtain mass spectra using a ToF-based SIMS instrument. A short pulse from the primary ion beam irradiates a spot and sputters secondary ions. The secondary ions are collected into the analyser and separated according to their (*m*/*z*) ratio. The process is repeated for the neighbouring pixels until a specified area (*x*–*y*) has been analysed with a defined number of pulses. Mass spectra are produced as the detector collects the sputtered secondary ions.

being identified. This can provide a fingerprint for the peak assignment, as the isotope distribution and peak intensity are fixed by the natural abundance.

An example of both positive and negative mass spectra obtained from the same material, an organic single crystal of rubrene, is shown in figure 3.14. Rubrene has the molecular formula $C_{42}H_{28}$ and a molecular mass of 532 amu and is composed of a collection of benzene rings. Both (a) positive and (b) negative secondary ion mass spectra are shown.

Rubrene has a relatively simple structure composed of only C and H atoms, but the mass spectra highlight the large differences that are observed between positive and negative secondary ionisation. In the positive mass spectrum (a), the parent molecule of $C_{42}H_{28}$ is clearly visible along with its isotopes. The molecular structure of the sample does not appear to have fragmented massively, and a series of smaller peaks (C_3H_3, C_4H_3, C_5H_3, C_6H_3) is visible at the lower mass end. In contrast, the negative mass spectrum (b) shows very intense peaks for the lower-order fragments (C_2H, C_4H, C_6H, C_8H and $C_{10}H$), which far overshadow any signal representing the parent molecule. Without some prior knowledge of the sample, identifying it from the negative spectrum alone would be difficult.

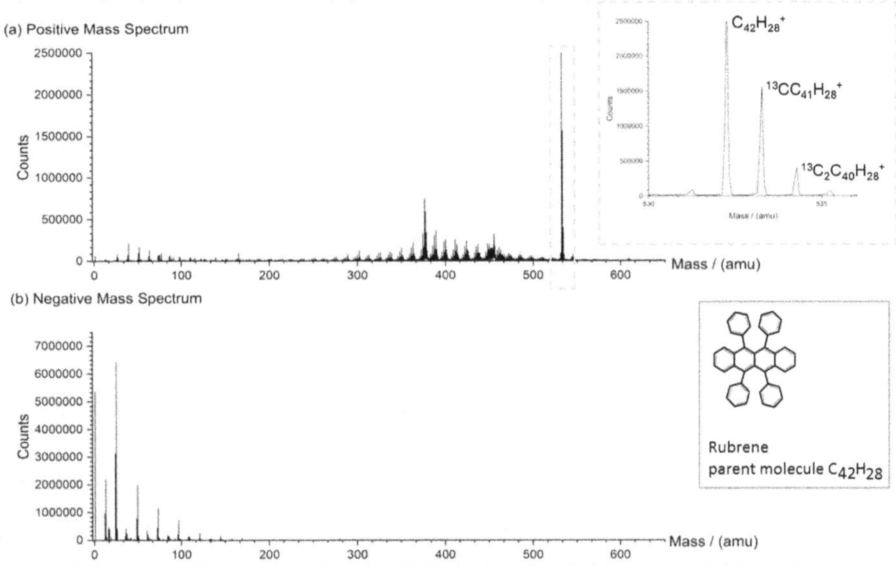

Figure 3.14. Mass spectra from the single organic crystal rubrene ($C_{42}H_{28}$): (a) a positive mass spectrum with isotopes of the parent molecule and (b) a negative mass spectrum.

3.3.1.2 Imaging/mapping

Ion imaging or mapping in SIMS shows the secondary ion intensities over the scanned area of interest. In this mode, we can obtain chemico-spatial information from the sample. For SIMS instruments with sequential mass analysers (quadrupole, magnetic sector), a preselection of ions is needed to generate the ion maps, whereas for time-of-flight systems with parallel detection, this is not necessary. Images are generated by selecting species of interest from the mass spectrum (see figure 3.15(b)). As the ion beam scans the sample, generating a mass spectrum, specific peaks may be selected, and as the precise spatial coordinates of the locations where the secondary ions were generated are known over the *x*–*y* plane, an ion map for the specifically selected masses can be created (see figure 3.15(c)).

To obtain laterally resolved ion images, the ion beam spot size should ideally be small, and a suitable number of pulses needs to be selected to ensure that there is beam overlap during the scanning of the analytical area. This is necessary to maintain continuous ion data from the scanned area; otherwise, the beam would irradiate individual spots within the scan area (see figure 3.16(b)), producing gaps in the ion image as regions of the scan area are left unanalysed. Again, for pulsed systems, this can be more of an issue; therefore, knowledge of the ion beam spot size is needed. For example, for an ion beam scanning an area of 100 μm^2 with 128 × 128 ion beam pulses (or pixels), the pulse resolution is ∼780 nm/pulse. A beam spot size of ∼1 μm is therefore needed for ion mapping. The table in figure 3.16 lists the nanometres per pulse for different numbers of ion beam pulses per scan in an area of 100 μm^2, thus indicating the minimum ion beam spot size required.

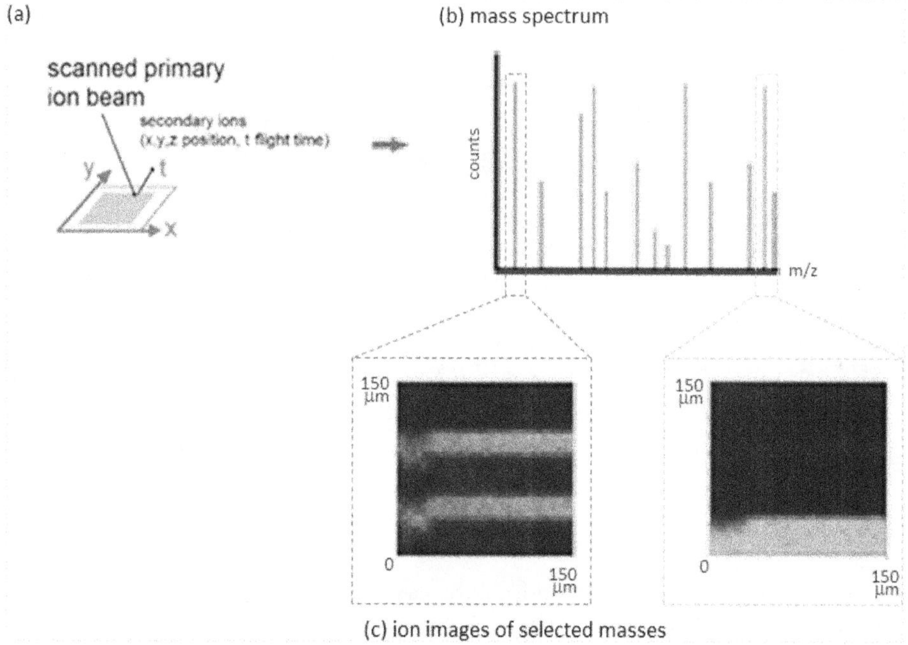

Figure 3.15. Steps needed to obtain ion images using a ToF-based SIMS instrument. (a) The sample is scanned. (b) A mass spectrum is acquired and specific peaks are selected from the mass spectrum. (c) The spatial distribution of the selected peaks is used to create ion images, as the spatial coordinates for the secondary ion signals are known.

The quality of the ion image is also influenced by the surface finish of the sample. A high-quality surface finish is essential for ion imaging, that is to say, a surface as smooth as possible. If the sample is an inorganic ceramic, for example, the surface should be polished to a mirror finish, as any roughness may result in blurred features of interest; in addition, topography may also create edge effects, distorting the secondary ion signal. Finally, if the spot size is small, secondary ion counts may be diminished, so binning an image can improve the counting statistics; however, this comes at the cost of reducing the resolution of the ion image.

Ion image data can be presented directly as the individual ion counts obtained over the scanned area, but more typically the ion images of interest are normalised to the total counts of the analysis, the count at each pixel being divided by the total ion count at that pixel. Normalisation helps to remove interference due to surface topography, instrument sensitivity, or matrix effects and helps to make the differences between the chemical species clearer. The simplest way to visually observe the distribution of different species and identify whether a species is correlated or anti-correlated is with an RGB overlay (figure 3.17).

A more statistical approach, such as colocalization analysis, can identify the degree of spatial correlation of two or more species. This type of analysis is a typical way to analyse image data collected by techniques such as electron microscopy. There are different ways to perform this kind of analysis; for example, it is possible

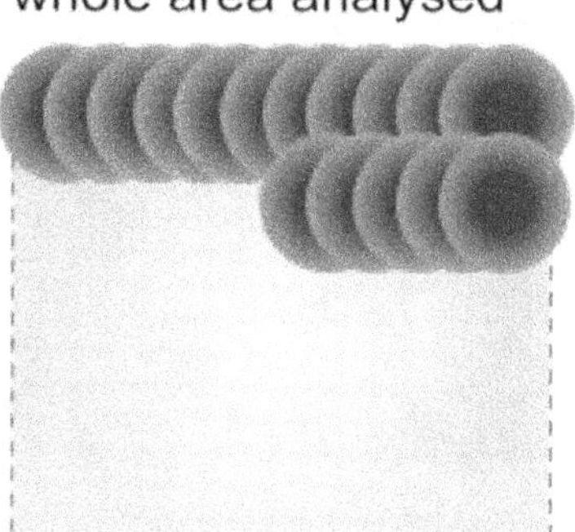

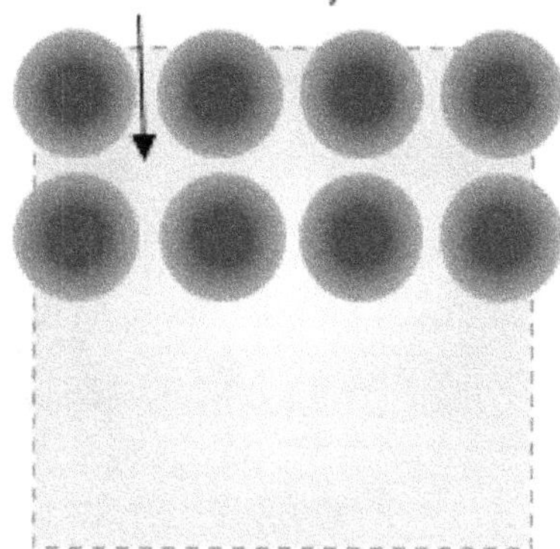

No. of pulses per scan	Pulse resolution* (nm/pulse)
128×128	780
256×256	390
512×512	195

*in 100 m^2

Figure 3.16. (a) Ion beam overlap in the *x*–*y* scan region compared to (b) no ion beam overlap in SIMS ion mapping, which leaves regions unanalysed. The table highlights the pulse (or pixel) resolution in 100 μm^2 for different numbers of pulses per scan.

to evaluate the intensity of a pixel in one channel against the intensity of a corresponding pixel in the other channel by obtaining a scatter plot and the colocalization coefficients, such as Pearson's or Spearman's correlation coefficient; alternatively, Manders' split coefficients can be calculated [57]. Generally, a correlation coefficient of $r = 1$ presents a perfect positive correlation, $r = 0$ presents very weak or zero correlation, while $r = -1$ presents a perfect negative correlation. The scatter plot can be simply obtained with Excel, image processing software such as Image J with JAcoP [58] and Colocalization Finder plugins [59], or MATLAB, Python, etc.

3.3.2 Dynamic mode

3.3.2.1 Depth profiling

For many systems, it is desirable to gain an understanding of how composition varies with depth, for example, to characterise interfaces, compositional changes through a material due to a processing step, and features beyond the surface region

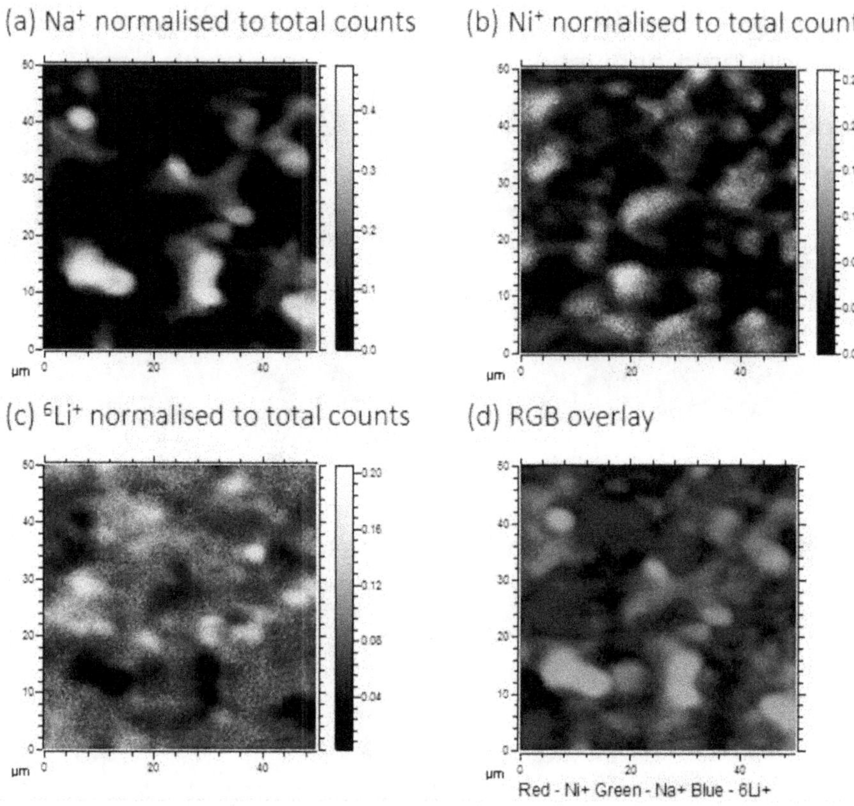

Figure 3.17. An example of normalised secondary ion maps of a sodium–lithium–nickelate sample: (a) sodium (Na+); (b) nickel (Ni+); (c) lithium (Li+); and (d) an RGB overlay of Ni+, Na+, and Li+.

[19]. In the dynamic mode, the ion beam dose exceeds the static limit, and layers of material are 'peeled away' to expose further layers of atoms or molecules for analysis. As the total ion beam dose increases, more and more material is removed, and a sputter crater is formed. This type of analysis is known as depth profiling SIMS.

The steps of depth profiling depend on the instrument being used. For SIMS instruments based on magnetic sector or quadrupole mass analysers, the primary ion beam is continuous when rastering across the target material (see figure 3.18(a)). Preselected secondary ions are collected, often limiting the collection to ~10 mass channels, and presented in a plot of secondary ion counts vs. sputter time. Once the analysis is finished, the final sputter crater can be measured *ex situ*, and the sputter time converted to a depth scale.

For instruments with ToF analysers, depth profiling requires a second ion beam (dual-beam instruments) for sputtering the material, as the primary ion gun is operated at very low currents and does not erode the sample. The first step in depth profiling is to raster the primary ion beam over the predefined analytical area (typically a square with a side length between 250 and 500 μm; see figure 3.18(b)). All

(a) Continuous ion beam irradiation (magnetic sector and quadrupole-based instruments)

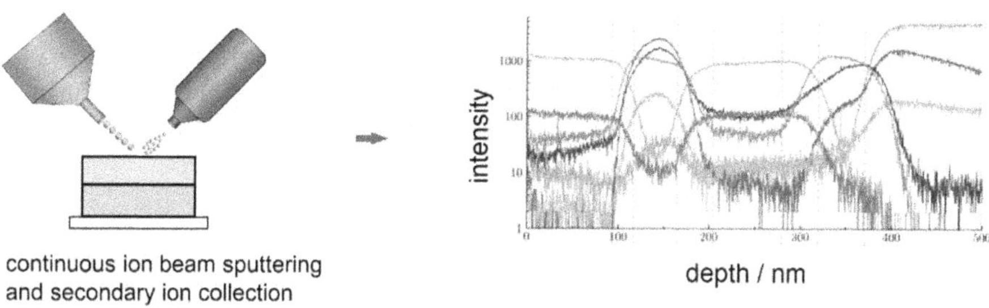

(b) Pulsing ion beam irradiation (time of flight-based instruments)

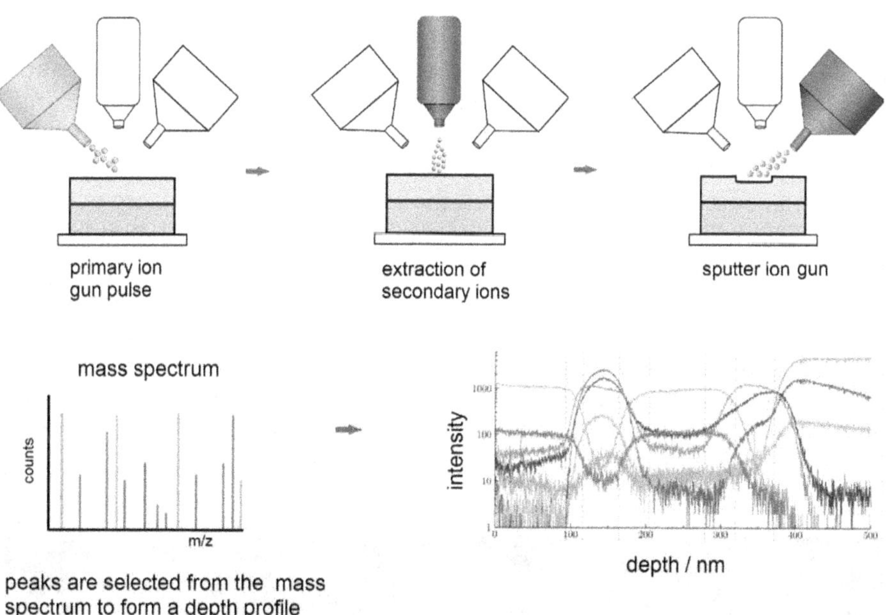

Figure 3.18. Schematic representation of the three main steps used in depth profiling with a dual-beam ToF-SIMS analyser. (a) The primary analytical ion beam is applied to the selected analysis area. (b) The secondary ions generated by the primary ion bombardment beam are collected and accelerated into the analyser. (c) A high-energy sputter beam is then rastered over the primary analysis region. (d) By selecting certain peaks, a depth profile of the species can be obtained (e).

ion beams are blanked while the secondary ions are selected. The sputter ion beam is then activated to remove material in a controlled fashion. The sputter ion beam current is typically on the order of tens to hundreds of nanoamps, and the defined sputter region must be greater than the analytical region to minimise any interference from material on the edge of the sputter crater, referred to as the 'edge effect'. The sequence of 'primary ion beam, secondary ion collection, sputter ion beam' is repeated sequentially until the desired depth of the sample has been measured or the feature of interest is reached. Peaks of interest are selected from the mass spectra generated by the primary ion beam, and a depth profile of those species is generated. Finally, to avoid any charge build-up that may occur on an insulating sample, a flood of low-energy electrons can be irradiated onto the surface. The current must be stable to maintain a constant sputter rate during the analysis.

3.3.2.3 3D data

As data is collected over the three planes, *x*, *y*, and *z*, three-dimensional data analyses can be obtained; however, to generate clearly resolved images, it is essential to acquire adequate secondary ion counts during the SIMS analysis. We can make an estimation of the number of counts that may be obtained for each voxel. For example, a depth profile is made over an area of 100 μm × 100 μm (0.01 cm^2) to a depth of 1 μm. During the analysis, 1000 × 1000 pixels scan the area, and 500 scans are made. The number of voxels in the analysis is therefore (1000 × 1000 × 500) 5×10^8. If the atomic areal density of the material is 10^{15} atoms cm^{-2}, there are 10^{13} atoms in each plane. If the material has an atomic spacing of 2 Å, there are 5000 planes in 1 μm. Therefore, the number of atoms in the available analytical volume is 5×10^{16}, which means there are $\sim 10^8$ atoms per voxel. If the species of interest is present at a concentration of 10%, then the number of atoms reduces to 10^7 atoms per pixel. This will be reduced further, as only a very small number of atoms become ionised; assuming a probability of ionisation of between 10^{-4} and 10^{-3}, the potential secondary ion counts obtained would be between 10 and 100 per voxel, respectively. Data density can be improved by binning the data, but again, this sacrifices the resolution of the final image.

The advantage of 3D analysis is, however, the ability to also see features on the *y–z* plane, which would otherwise be hidden in a 2D depth profile. Figure 3.19 shows a phase laser deposited yttria-stabilised zirconia (YSZ) film grown on a strontium

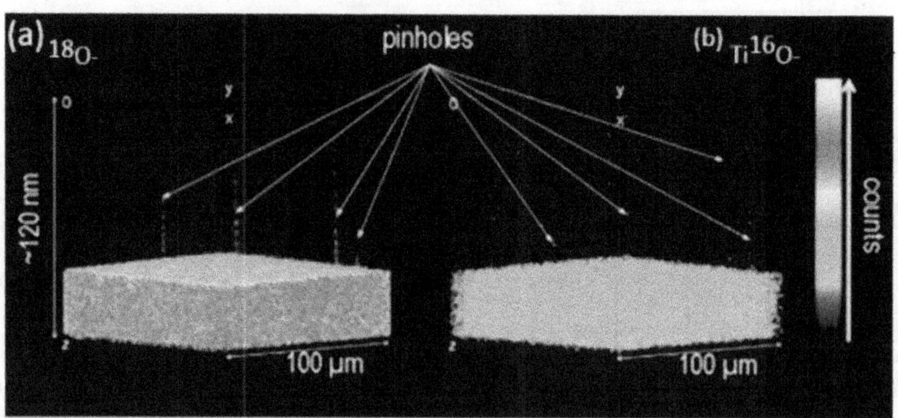

Figure 3.19. A three-dimensional ion images of: (a) $^{18}O^-$ through pinholes in a YSZ thin film on an STO substrate and (b) the $Ti^{16}O^-$ signal. Reprinted with permission from [60], Copyright (2013), with permission from Elsevier.

titanate (STO) substrate [60]. The sample has been exposed to an ^{18}O atmosphere and then depth profiled. The 3D image, figure 3.19(a), shows that pinholes of micron size are visible throughout the YSZ film, and (b) the $Ti^{16}O^{-}$ signal indicates that TiO^{-} has been sputtered from the substrate through the pinholes. It would be extremely difficult to pick up such features using an electron microscope, for example. As most of the pinholes do not run all the way to the surface, these would not necessarily have been detected with just an elemental surface map.

References

[1] Storms H A, Brown K F and Stein J D 1977 *Anal. Chem.* **49** 2023

[2] Benninghoven A 1976 *Crit. Rev. Solid State Sci.* **6** 291

[3] Magee C W 1981 *Nucl. Inst. Methods* **191** 297

[4] Kollmer F 2004 *Appl. Surf. Sci.* **231–2** 153

[5] Toubol D, Kollmer F, Niehus E, Brunelle A and Laprevote. O 2005 *J. Am. Soc. Mass. Spectrom.* **16** 1608

[6] Bayly A R, Waugh A R and Anderson K 1983 *Nucl. Instr. Methods. Phys. Res.* **218** 375

[7] Levi-Setti R, Wang Y L and Crow G 1986 *App. Surf. Sci.* **26** 249

[8] Ninomiya S, Ichiki K, Yamada H, Nakata Y, Seki T, Aoki T and Matsuo J 2009 *Rapid Commun. Mass Spectrom.* **23** 1601

[9] Bich C, Havelund R, Moellers R, Touboul D, Kollmer F, Niehus E, Gilmore I S and Brunelle A 2013 *Anal. Chem.* **85** 7745

[10] Delcorte A, Cristaudo V, Zarshenas M, Merche D, Reniers F and Bertrand P 2015 *Plasma Process. Poly.* **12** 905

[11] Winograd N 2018 *Annu. Rev. Anal. Chem.* **11** 29

[12] Ziegler J F Interactions of Ions with Matter www.SRIM.org

[13] Vickerman J C and Gilmore I S 2009 *Surface Analysis the Principal Techniques* 2nd edn (New York: Wiley)

[14] Clegg J B 1991 *Growth and Characterisation of Semiconductors* ed R A Stradling and P C Klipstein (London: Adam Hilger)

[15] Benninghoven A, Rüdenauer F G and Werner H W 1987 *Secondary Ion Mass Spectrometry—Basic Concepts, Instrumental Aspects, Application and Trends* (New York: Wiley)

[16] Ziegler J F and Biersack J P 1985 *The stopping Range of Ions in Matter* (Pergammon Press)

[17] Winters H F and Coburn J W 1976 *Appl. Phys. Lett.* **28** 176

[18] Bolbach G, Viari A, Galera R, Brunot A and Blais J C 1992 *Int. J. Mass Spectrom. Ion Process.* **112** 9

[19] Wilson R G, Stevie F A and Magee C W 1989 *Secondary Ion Mass Spectrometry: A Practical Handbook for Depth Profiling and Bulk Impurity Analysis* (New York: Wiley)

[20] Liau Z L, Tsaur B Y and Mayer J W 1979 *J. Vac. Sci. Technol.* **16** 121

[21] Andersen H H 1979 *Appl. Phys.* **18** 131

[22] Littmark U and Hofer W O 1980 *Nucl. Instr. Meth.* **168** 329

[23] Williams P 1980 *Appl. Phys. Lett.* **36** 758

[24] Williams P and Baker J E 1981 *Nucl. Instr. Meth.* **182–3** 15

[25] Magee C W, Honig R E and Evans C A Jr 1982 *SIMS III* ed A Benninghoven, J Giber, J Laszlo, M Riedel and H W Werner (Berlin: Springer) p 172

[26] Liu R, Ng C M and Wee A T S 2003 *App. Surf. Sci.* **203–4** 256–9

[27] Wehner G K and Hajicek D J 1971 *J. Appl. Phys.* **42** 1145

[28] Hofer W O and Liebl H 1975 *Appl. Phys.* **8** 359

[29] Tompkins H G 1987 *Surf. Int. Anal.* **10** 105

[30] McPhail D S, Dowsett M G, Fox H, Houghton R, Leong W Y, Parker E H C and Patel G K 1988 *Surf. Int. Anal.* **11** 80

[31] Deline V R, Reuter W and Kelley R 1986 *SIMS V* ed A Benninghoven, R J Colton, D S Simons and H W Werner (Berlin: Springer) p 299

[32] Weast R C 1987 *Handbook of Chemistry and Physics* 67th edn (Boca Raton, FL: CRC Press)

[33] Hues S M and Williams P 1986 *Nucl. Instr. Meth. Phys. Res.* B **15** 206

[34] Homma Y and Wittmaack K 1990 *Appl. Phys.* A **50** 417

[35] Boudewijn P R and Vriezema C J 1988 *SIMS VI* ed A Benninghoven, A M Huber and H W Werner (New York: Wiley) p 499

[36] Vriezema C J, Janssen T F and Boudewijn P R 1989 *Appl. Phys. Lett.* **54** 198

[37] Maier M, Bimberg D, Baumgart H and Phillip F 1982 *SIMS III* ed A Benninghoven, J Giber, J Laszlo, M Riedel and H W Werner (Berlin: Springer) p 336

[38] Vandervost W, Remmerie J, Shepherd F R and Swanson M L 1986 *SIMS V* ed A Benninghoven, R J Colton, D S Simons and H W Werner (Berlin: Springer) p 288

[39] Jones E A, Fletcher S J, Thompson C E, Jackson D A, Lockyer N P and Vickerman J C 2006 *Appl. Surf. Sci.* **252** 6844–54

[40] Jones E A, Lockyer N P and Vickerman J C 2007 *Int. J. Mass Spec.* **260** 146–57

[41] Postawa Z, Czerwinski B, Szewczyk M, Smiley E J, Winograd N and Garrison B J 2003 *Anal. Chem.* **75** 4402

[42] Postawa Z, Czerwinski B, Szewczyk M, Smiley E J, Winograd N and Garrison B J 2004 *J. Phys. Chem.* B **108** 7831

[43] Delcorte A, Wehbe N, Bertrand P and Garrison B 2008 *Appl. Surf. Sci.* **255** 1229

[44] Declorte A, Garrision B J and Hamraoui K 2009 *Anal. Chem.* **81** 6676–686

[45] Russo M F Jr, Wojciechowski I A and Garrison B J 2006 *Appl. Surf. Sci.* **252** 6423–5

[46] Bich C, Havelund R, Moellers R, Touboul D, Kollmer F, Niehuis E, Gilmore I S and Brunelle A 2013 *Anal. Chem.* **85** 7745–52

[47] Mouhib T, Poleuins C, Wehbe N, Michels J J, Galagan Y, Houssiau L, Bertrand P and Delcorte A 2013 *Analyst* **138** 6801–10

[48] Aoyagi S, Fletcher J S, Sheraz (Rabbani) S, Kawashima T, Berrueta-Razo I, Henderson A, Lockyer N P and Vickerman J C 2013 *Anal. Bioanal. Chem.* **405** 6621–8

[49] Shard A G *et al* 2012 *Anal. Chem.* **84** 7865–73

[50] Matsuo J, Okubo C, Seki T, Aoki T, Toyoda N and Yamada I 2004 *Nucl. Instrum. Methods Phys. Rev.* B **219–20** 463–7

[51] Czerwinski B and Delcorte A 2013 *J. Phys. Chem.* C **117** 3595–604

[52] Postawa Z, Paruch R, Rzeznik L and Garrison B J 2013 *Surf. Int. Anal.* **45** 35–8

[53] Kayser S, Rading D, Moellers R, Kollmer F and Niehuis 2013 *Surf. Int. Anal.* **45** 131–3

[54] Matsuo J, Toyoda N, Akizuki M and Yamada I 1997 *Nucl. Instru. Methods Phys. Res.* B **121** 459–63

[55] Seki T, Murase T and Matsuo J 2006 *Nucl. Instru. Methods Phys. Res.* B **242** 179–81

[56] Seah M 2013 *J. Phys. Chem.* C **117** 12622–32

[57] Schober P, Boer C and Schwarte L A 2018 *Anesth. Analg.* **126** 1763

[58] Bolte S and Cordelieres F P 2006 *J. Micro* **224** 213–32

[59] Laummonerie C, Mutterer and Carl P http://questpharma.u-strasbg.fr/html/colocalization-finder.html Colocalization_Finder

[60] Stender D, Cook S, Kilner J A K, Dobeli M, Conder K, Lippert T and Wakaun A 2013 *Solid State Ionics* **249–50** 56–62

IOP Publishing

Secondary Ion Mass Spectrometry and Its Application to Materials Science (Second Edition)

Sarah Fearn

Chapter 4

Data handling

4.1 Quantification based on relative sensitivity factors and ion implant standards

In the basic SIMS equation, the concentration C_x of an impurity species x in a matrix is proportional to the secondary ion current of x being collected:

$$I_s^x = I_p C_x S \gamma F, \tag{4.1}$$

where I_s^x is the collected secondary ion current, I_p is the primary ion beam current, C_x is the concentration of impurity x, S is the sputter yield of x, γ is the ionisation efficiency of x, and F is the instrument transmission.

It therefore seems straightforward that from the secondary ion counts that are collected of a specific species from a sample, the counts can easily be converted and quantified into concentration. Unfortunately, the SIMS secondary ion count is not self-quantitative due to the matrix effect and the variability of both the sputter yield S and the ionisation efficiency γ. One area of material analysis that can be successfully quantified is in the area of depth profiling. The quantification of certain depth profiles can be achieved using relative sensitivity factors (RSFs) or ion implant standards.

RSFs are obtained from ion implant or bulk-doped samples. The RSF is used to convert secondary ion intensity into concentration:

$$Cx/(\text{atoms/cm}^3) = \text{RSF}\frac{Ix}{Im}, \tag{4.2}$$

where I_m is the secondary ion signal of the matrix containing the implant species x, and I_x is the secondary ion count of species x.

A collection of RSFs has been built up over the years, in particular for the semiconductor industry [1], as quantification of implant profiles is essential for the

doi:10.1088/978-0-7503-3331-3ch4

Table 4.1. Selected RSFs for 28 elements implanted into silicon and measured using Cameca SIMS instruments (RSF units cm^{-3}) [2].

Ion species	Beam	RSF	Ion species	Beam	RSF
H+	O	6.2×10^{24}	Co+	O	5.3×10^{22}
H−	Cs	4.8×10^{21}	Co−	Cs	2.0×10^{24}
Li+	O	5.9×10^{20}	Zr+	O	2.4×10^{21}
Li−	Cs	5.9×10^{24}	Zr−	Cs	1.0×10^{25}
B+	O	6.5×10^{22}	Mo+	O	2.3×10^{22}
B−	Cs	2.7×10^{27}	Mo−	Cs	2.0×10^{25}
O+	O	7.9×10^{25}	In+	O	1.5×10^{21}
O−	Cs	2.4×10^{22}	In−	Cs	1.8×10^{26}
Al+	O	1.4×10^{21}	Sn+	O	3.0×10^{22}
Al−	Cs	1.5×10^{25}	Sn−	Cs	1.8×10^{23}
P+	O	1.1×10^{24}	Ba+	O	1.5×10^{21}
P−	Cs	1.2×10^{23}	Ba−	Cs	1.6×10^{26}
Fe+	O	2.7×10^{22}	La+	O	2.8×10^{21}
Fe−	Cs	5.3×10^{25}	La−	Cs	8.6×10^{24}

development of doped devices. Typically, the RSFs for As, P, and B in silicon, gallium arsenide (GaAs), and indium gallium arsenide (InGaAs) can be found in the literature [2] (see table 4.1 for RSFs of implants in Si).

Ion implant standards can also be fabricated, and these provide accurate quantification for a known impurity species in a matrix. This approach is highly flexible, as any element or isotope of an element can be implanted into a matrix material, and the energy of the implant may be varied to fix the depth of the implant peak as required. The implant dose may also be varied, thus changing the concentration of the implant peak. The substrate (matrix) of the implant can also be selected as required, and indeed, implanting into structures can also be carried out.

4.1.1 Steps needed to quantify a sample using an ion implant standard

1. An ion implant standard needs to be made. This standard should relate to the sample(s) that will subsequently be quantified. For example, if the samples to be quantified have a shallow implant profile, such as boron in silicon, the ion implant standard should be of boron in silicon, implanted with a known implant dose at an implant energy that produces an implant profile in boron similar to the samples to be tested. This ensures that the same experimental SIMS conditions (or conditions as close as possible) can be used for measuring both the standard and the sample. This helps to reduce errors between the two measurements.
2. The ion implant standard is then depth profiled using SIMS, and the RSF is then calculated from this data. Figure 4.1(a) shows an example of the raw

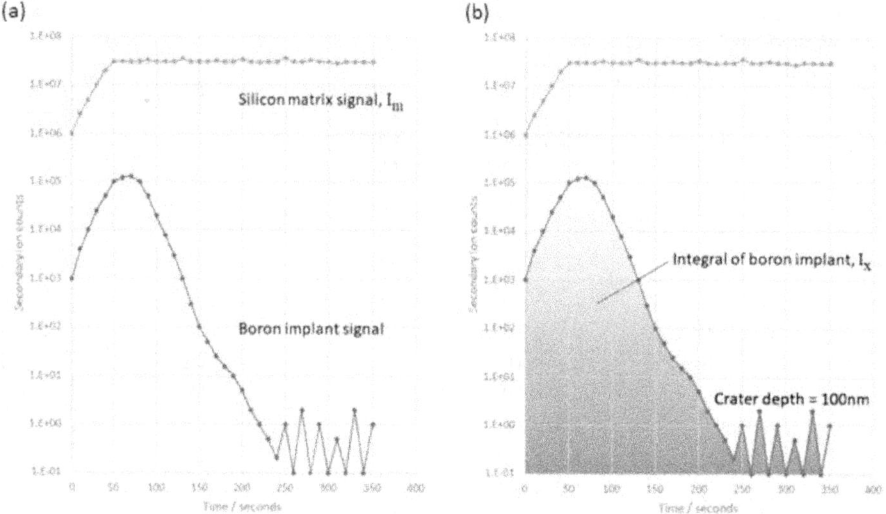

Figure 4.1. Boron ion implant at a dose of 1×10^{15} ions cm^{-2}. (a) A schematic of the SIMS boron depth profile of the ion implant and (b) the integral of the implant and final crater depth of the analysis.

data obtained from depth profiling a boron ion implant standard. The dose of the implant is 1×10^{15} ions cm^{-2}.

3. From the raw data of the SIMS depth profile, the average secondary ion signal for the Si matrix, $\boldsymbol{I_m}$, is measured. Figure 4.1(a) gives a value of I_m, $= \sim 5 \times 10^7$ ions.
4. The average secondary ion signal, $\boldsymbol{Ix}$, for the **B** implant is measured. This is the integral of the profile, highlighted in blue in figure 4.1(b), divided by the analysis time:

$$Ix = \frac{3.7 \times 10^6}{300} = 1.2 \times 10^4 \text{ ions.}$$

5. The average ion implant concentration $\boldsymbol{C_x}$ is calculated using the known implant dose and the depth of the crater formed in the SIMS measurement, shown as 100 nm in figure 4.1(b):

$$Cx = \frac{\text{implant dose}}{\text{crater depth}} = \frac{1 \times 10^{15}}{1.0 \times 10^{-4}} = 1 \times 10^{19} \text{ ions/cm.}$$

6. The RSF for the implant profile can now be calculated by rearranging equation (4.2) and inputting the values obtained in steps 3, 4, and 5, to give:

$$\text{RSF} = Cx\frac{Im}{Ix} = (1 \times 10^{19})\frac{5 \times 10^7}{1.2 \times 10^4} = 0.42 \times 10^{23}.$$

7. The calculated RSF from the ion implant can then be applied to the SIMS data from the samples to be quantified to convert the y-axis from secondary ion counts to concentration using equation (4.2).

4.2 Quantification without standards

4.2.1 Using a secondary technique

Relative sensitivity factors for semiconductors such as silicon are well established, and implanting into silicon or other semiconductors is relatively easy. However, many samples cannot be easily quantified because their matrices are more complex, making it challenging or too expensive to create suitable standards. For example, glass is an amorphous material with a complicated matrix, which means that any variation in the bulk matrix material will lead to changes in S and γ of any species being detected from the matrix. In this case, a very large number of standards would be needed to accommodate the varying S and γ values. An alternative way to quantify more complex materials, such as glass, is to combine the SIMS analysis with a second technique that can quantify the bulk chemistry of the sample.

Figure 4.2(a) is a SIMS depth profile of a soda lime glass, showing some of the main components of the glass, Na, Al, Si, and Ca, and how their distribution changes from the surface towards the bulk of the sample. The x-axis has been converted to depth, and the profile shows that in the top 20 nm of the sample, the Si ions are enhanced and the Na ions are depleted compared to their respective bulk levels. This depletion at the near surface can lead to problems in glass manufacture, so it would be beneficial to calculate the concentration of leached sodium ions as a way of estimating the diffusion characteristics of the glass and particularly the loss of

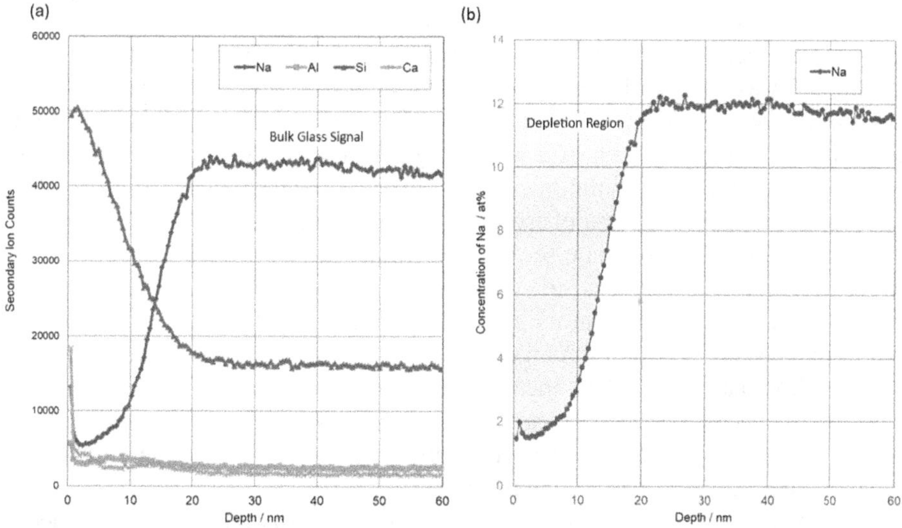

Figure 4.2. (a) A SIMS depth profile of a soda lime glass sample. The near surface shows extensive leaching of sodium ions. (b) The highlighted depletion region indicates the amount of sodium ions leached from the glass in the depletion region.

Na ions at different humidities and temperatures [3, 4]. To do this, however, the *y*-axis needs to be converted from secondary ion counts to concentration.

From the basic SIMS equation, we know that the measured secondary ion counts for an individual species are proportional to the concentration of that species; i.e. the secondary ion signal of Na is proportional to the concentration of Na in the sample, $I_s^{Na} \propto C_{Na}$. If we use a secondary technique such as electron probe microanalysis (EPMA), we can ascertain the bulk concentrations of all the elements in the glass, which can then be used to convert the *y*-axis into concentration. As the EPMA measurement gives the Na ion concentration in the bulk of the glass, we can equate this value to the average Na SIMS signal from the bulk and calculate a conversion factor from the ratio of $\frac{\text{(bulk concentration of Na from EPMA)/(atomic\%)}}{\text{(average Na SIMS count in bulk)/(counts)}}$. This can then be used to convert and back-calibrate the secondary ion signal for the whole profile to concentration (see figure 4.2(b)). The *y*-axis is then converted from secondary ion counts to concentration in atomic %. With the *y*-axis now converted, by integrating the depletion region, it is possible to calculate the number of sodium ions leached from the surface. This can also be repeated for the other SIMS signals if required.

By adding a secondary analytical technique to obtain the bulk concentration level, we can overcome some of the issues associated with the matrix effect for the amorphous material, quantify the SIMS data, and exploit one of the unique strengths of the SIMS technique: the ability to show how the chemistry of the material changes over very small distances. In this case, the distance involved is the top 50 nm of the sample surface, and the data is obtained at high resolution.

4.2.2 Using isotopes

A unique strength of the SIMS technique is its ability to accurately and precisely measure isotopes. The application of isotopes as a way to unambiguously identify processes that occur in materials is one that has been applied to ceramics, metals, and alloys, as well as the biosciences. In this section, the focus is primarily on the diffusion of oxygen through ceramic oxide materials, as the experimental procedure nicely highlights how, as part of a well-designed experiment, the strength of SIMS can be maximised while minimising its artefacts—specifically, matrix effects. Information regarding the use of isotopes in metals and alloys and the biosciences can be found in the appropriate sections of chapter 5.

Determining the oxygen diffusion coefficients of ceramic oxides is essential for their applications as oxygen separators and solid oxide fuel cells (SOFCs) [5, 6]. By performing oxygen isotope exchange experiments (where ^{16}O is basically replaced by ^{18}O), SIMS can be used to measure the subsequent oxygen signals within the exchanged materials. The SIMS measurement neatly avoids the pitfalls and difficulties associated with matrix effects that arise in SIMS analyses, as the ratios of the chemically identical isotopes are measured.

To obtain oxygen diffusion coefficients, D^*, (and the surface exchange coefficient k^*) of a bulk ceramic, an isotope exchange diffusion experiment is performed. The experiment involves initially equilibrating a bulk ceramic material in an oxygen 16 (^{16}O) atmosphere at a specific temperature (typically related to the operational temperature of the ceramic). Once equilibrated, the atmosphere is then changed to one that is enriched with the oxygen 18 isotope (^{18}O). The sample is then heated to the same temperature as that used in the first step, whereby the ^{18}O diffuses into the sample, replacing the ^{16}O, as there is a chemical imbalance in the sample between the oxygen isotopes, highlighted in the schematic of figure 4.3(a). After a fixed period of time, the sample is quenched to 'lock in' the ^{18}O diffusion profile. Once the sample has cooled, there are a number of ways the ^{18}O diffusion profile can be measured via SIMS, depending on how far the profile extends into the sample.

If the diffusion profile is less than ~2 μm, basic depth profiling can be carried out to obtain the secondary ion signals of the ^{16}O and ^{18}O isotopes. Diffusion profiles are typically greater than this, as diffusion profiles can extend up to several hundred microns depending on the specific material. For these longer profiles, it is necessary to prepare a cross section of the sample (see figure 4.3(b)). SIMS ion mapping or line scans can then be measured over the polished cross section.

Figure 4.4 highlights the $^{18}O^-$ SIMS ion map across the cross section. Ion maps of both $^{16}O^-$ and $^{18}O^-$ are obtained, and the matrix of data from the ion map is then integrated into a column of data for each isotope [7].The secondary ion counts I^{18O-} and I^{16O-} are then used to calculate the ^{18}O isotopic fraction of the diffusion profile, $C_x = \left(\frac{I^{18O-}}{I^{18O-} + I^{16O-}} \right)$.

The ^{18}O isotopic fraction is then combined with C_g, the isotopic fraction of ^{18}O in the gas used in the ion exchange step, C_{bg}, the ^{18}O background isotopic fraction, and t, the ^{18}O anneal time, to plot $C'(O)$, the ^{18}O isotope fraction profile *vs*. distance x (the length of the line scan or ion map; see figure 4.5). The data is fitted to the solution of Fick's second law of diffusion for a semi-infinite medium, shown below in equation (4.3) [8], using a regression analysis fit in MATLAB [9] to identify D^* and k^*, i.e. the diffusion and surface exchange coefficients, respectively:

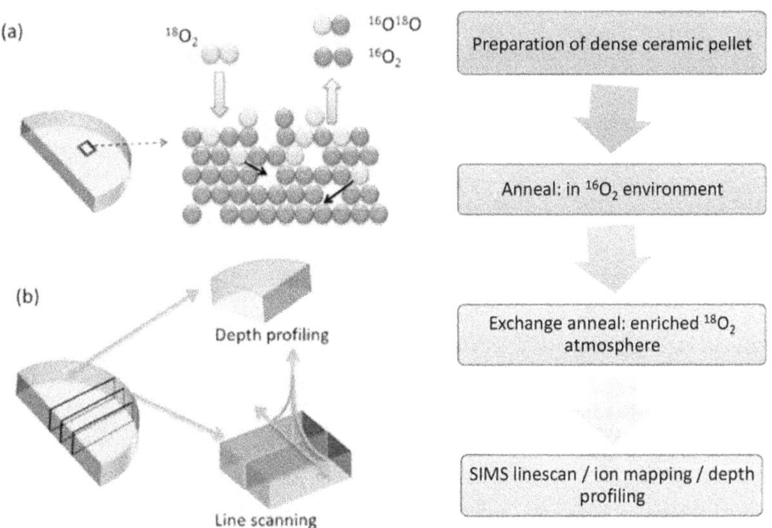

Figure 4.3. Schematic of the steps needed for the isotope exchange experiment.

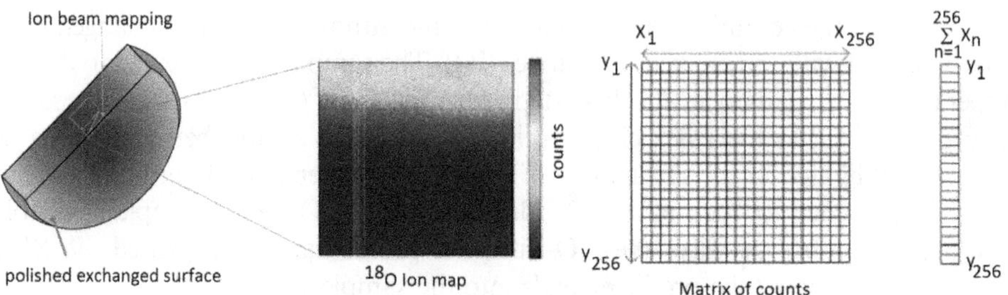

Figure 4.4. Schematic of SIMS ion mapping across the cross section of the isotope-exchanged pellet.

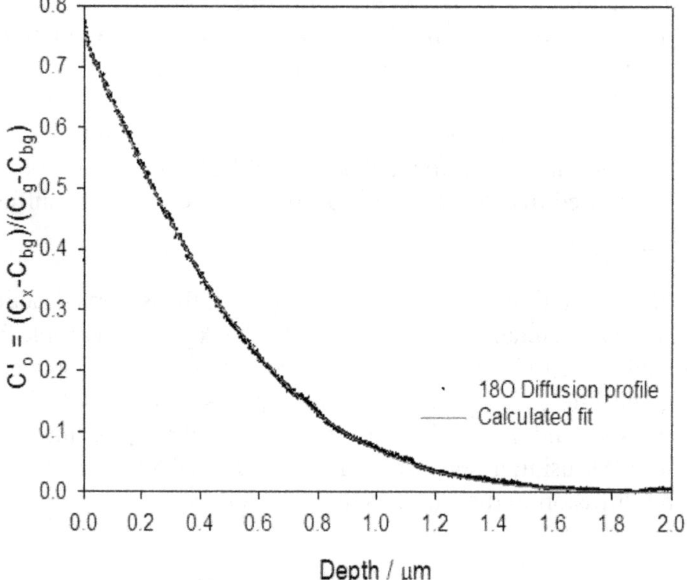

Figure 4.5. Plot of the ^{18}O isotopic fraction, $C'(O)$, vs. depth/distance with a calculated diffusion fit to equation (4.3).

$$C'(O) = \frac{C_x - C_{bg}}{C_g - C_{bg}} = \text{erfc}\left(\frac{x}{2\sqrt{D_* t}}\right) - \exp\left(hx + h^2 D_* t\right)\text{erfc}\left(\frac{x}{2\sqrt{D_* t}} + h\sqrt{D_* t}\right), \tag{4.3}$$

where $h = \frac{k_*}{D_*}$.

For bulk diffusion, transport parameters over many orders of magnitude can be obtained using the isotope exchange diffusion SIMS experiment. The range of diffusion coefficients that can be measured via this technique can vary from 10^{-5} to $10^{-19}\text{cm}^2\ \text{s}^{-1}$ [10].

4.3 Multivariate analytical techniques

SIMS data obtained from organic materials typically contains a vast amount of data. A full mass spectrum can be obtained at each pixel of a SIMS spectrum or image. The data includes hundreds, if not thousands, of peaks, some of which are more significant than others. The richness of the SIMS data is one of its great advantages as a technique, but it is not possible to manually analyse the vast array of data produced. To overcome this, a branch of chemometrics, multivariate analysis (MVA), has been widely adopted within SIMS to usefully analyse the large data sets acquired. In this section, a brief overview will be given of this subject. For more detailed information, there are many excellent reviews and academic texts available [11–13].

Chemometrics is commonly used within the biosciences to extract useful information from complex systems and is widely applied to the fields of, for example, spectroscopy and nuclear magnetic resonance (NMR). MVA, a core subset of chemometrics, is a set of statistical methods that can analyse large data sets. When applied to SIMS spectra or images, MVA can identify patterns in high-dimensional data (data sets with a large number of features such as columns relative to the number of observations, i.e. rows or data points). It can be used to reduce data sets from many hundreds to a few significant peaks to effectively help in pinpointing active ion species. For example, it is used in identifying the different polymer surfaces that display antimicrobial attachment or the impact of cleaning treatments on historical polymer-based artefacts [14, 15]. It can also distinguish and classify various cell and tissue types.

4.3.1 Basic multivariate analysis procedure

Data acquisition:	(a) Collect SIMS data, i.e. mass spectra or images, and export them to an ASCII text file.
Data preprocessing:	(a) Mass peak selection. Relevant peaks (ions) are preselected based on known chemistry or statistical importance. This can also be done by a peak-picking algorithm. (b) Normalisation. The data is normalised per pixel or per ion count to account for signal variation between scans. (c) Mean centering. The average value is subtracted from each variable.
Select an MVA method:	(a) Principal component analysis (PCA) is used to find major variance differences. (b) Partial least squares discriminant analysis (PLS-DA) is used to distinguish between different groups and classes. (c) Multivariate curve resolution (MCR) is used to clearly separate regions in images.

Various software packages exist that perform MVA on SIMS data, either using built-in analysis in vendor software or custom analysis through scripting in MATLAB or Python. Images can also be statistically analysed using ImageJ with MVA plugins.

MVA methods are extremely useful tools for better understanding large SIMS data sets, with the benefit of reducing these large data sets into a manageable set of variables. The use of MVA techniques does not, however, replace the need to understand the methodologies used to analyse SIMS data and to be aware of experimental artefacts such as the matrix effect. As data science continues to evolve, this will undoubtedly influence the methodologies used to analyse complex SIMS data sets.

4.4 Sample handling and preparation

With all sample handling and preparation for SIMS analysis, the aim is to minimise contamination from extraneous sources and prepare a sample that exploits the strength of the technique to its maximum. The ideal sample has a dense solid form with an atomically flat surface or a very well-polished surface (to a mirror finish). This ensures even sputtering through the sample if depth profiling is intended and reduces the artefacts that may be induced by surface roughness if highly resolved images are required. However, obtaining the ideal sample is often impractical or not possible and is most likely not representative of a sample that has been tested or cycled in a previous experimental setup; for example, samples retrieved from coin cell batteries. The sections below outline some general good practices and tips for obtaining useful samples for SIMS measurements, as a SIMS measurement is only as good as the sample it is analysing!

4.4.1 General good practice

Before SIMS measurements take place, care should be taken when storing samples to avoid surface contamination and/or degradation of the sample. Common surface contaminants, such as moisture, can lead to the formation of unwanted hydroxide or oxide surface layers, which in themselves can lead to further segregation of impurities and of the sample species of interest. To minimise this effect as much as possible, it is advised that samples are kept in lint-free paper and stored in a moisture-controlled desiccator. Polymer membrane boxes are to be avoided at all costs, as they deposit polymer layers onto samples, which are detected in mass spectra and can be difficult to remove.

Samples should always be handled with gloves as a matter of basic health and safety. Always handle the sample holder while wearing gloves to avoid the deposition of salts from the hands onto the holder. The holder must be kept clean, as it will regularly go in and out of the ultrahigh vacuum (UHV) main chamber.

Sample surfaces can be cleaned before SIMS analysis, ideally with just a very clean air blower to remove any dust or fibres. Surfaces can be carefully wiped with lint-free cotton and isopropanol (IPA). Ensure that the surface has fully dried before loading the sample into the vacuum system, which will draw off any organic residue.

If the sample is stable, a heat lamp may help remove some of the organic residue; however, the IPA is liable to leave behind a monolayer of organic residue on the surface. This will not be an issue in a depth profile over hundreds of nanometres, but it will be an issue for very near-surface analysis measurements.

Samples should be attached to a sample holder using clean metal clips, and the use of adhesive tape should be minimised and reserved for the mounting of the most difficult samples: those that are too small or unevenly shaped to fit under clips or holder windows. If tape is required, platinum tape can be used to attach samples to holders and also provides a useful conductive contact between the sample and the holder. Once well-stored and cleaned samples have been mounted, they can be carefully placed into the SIMS instrument.

4.4.2 Air-sensitive samples

Along with the general application of good sample handling procedures outlined above, for materials that are air-sensitive, such as many of the materials used in battery applications, storage in an argon-filled glove box is essential. As samples will be mounted within the glove box, it is helpful to prepare the holder (e.g. position sample clips) as much as possible outside the glove box, as it is easier to manipulate the fittings. Once the sample holder is prepared, the holder (and any other essential equipment) should be put into the glove box 24 h before samples are mounted.

For chemical analysis of the solid–electrolyte interface (SEI), storage conditions and sample handling are even stricter. Ideally, the glove box used for battery assembly and disassembly should be separate from the glove box used to hold samples for analysis. This is important to avoid cross-contamination of the organic electrolyte used to fabricate lithium and sodium batteries.

Air-sensitive samples, once mounted, need to be transferred from the glove box to the SIMS instrument using a suitable vacuum transfer system. Vacuum transfer systems typically function in one of two ways: either the sample is kept under a high dynamic vacuum throughout the whole transfer process, or the suitcase is kept at a low inert pressure after having been held in the glove box. For these suitcases, which are normally placed into the load lock, the load lock is initially pumped down to a mid-pressure vacuum ($\sim 10^{-5}$ mbar) before the suitcase is opened and the atmosphere inside is pumped away. The specific type of vacuum suitcase used depends upon the instrumentation available.

4.4.3 Biological sample preparation

The high vacuum demands of SIMS require samples to be UHV compatible. When preparing biological samples, therefore, careful consideration is needed to preserve the structural and chemical integrity of the samples. A basic overview of the steps required is as follows:

Fixing—preserve the structure and reduce the degradation of the biological features of interest.

Cryo sectioning—use a cryostat to create ultrathin cross sections for internal analysis.

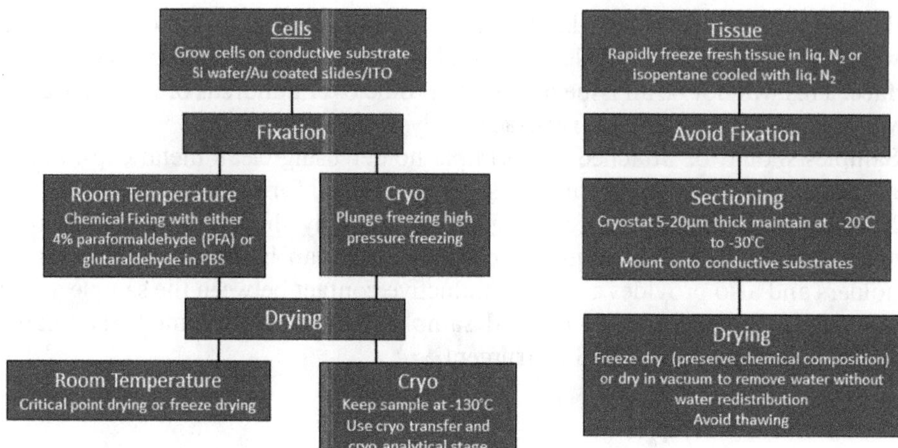

Figure 4.6. Workflows for the preparation of biological samples, namely cells and tissues, at room temperature.

Drying—biological material contains water that must be carefully removed for vacuum compatibility.

Mounting—place the sample on a conductive substrate such as a Si wafer, a Au-coated slide, or an indium tin oxide (ITO) slide.

Surface coatings—these should be avoided because they can affect the surface chemistry, which is something SIMS is very sensitive to.

Preparing cells or tissues for SIMS may require different sample preparation workflows (see figure 4.6), depending on the analytical goals: whether it is biological structure that is to be maintained, molecular distribution, or both. For example, lipid mapping in cells or tissue depends on maintaining chemical fidelity. In preparing a sample for SIMS analysis, care needs to be taken to preserve the native lipid distribution and prevent delocalisation, as lipids are easily redistributed by heat, solvents, or surface contamination.

Cryo-preparation is considered to be the 'gold standard' for biological samples. However, this may not be possible for various reasons; for example, SIMS may have not been considered a part of the experimental protocol before the samples were obtained. Cryo-stages are not consistently available, so alternative techniques are used with an understanding of the impact the preparation has on the data that can be obtained.

References

[1] Wilson R G 1988 *J. Appl. Phys.* **63** 5121

[2] Wislon R G and Novak S W 1991 *J. Appl. Phys.* **69** 466

[3] Fearn S, McPhail D S and Oakley V 2005 *Phys. Chem. Glasses* **46** 505–11

[4] Rodrigues A, Fearn S and Vilarigues M 2018 *Corros. Sci.* **145** 249–61

[5] Kilner J A, Steele B C H and Ilkov L 1984 *Solid State Ionics* **12** 89–97

[6] Carter S, Selcuk A, Chater R J, Kajda J, Kilner J A and Steele B C H 1992 *Solid State Ionics* **53–6** 597–605

[7] DeSouza R A, Zehnpfenning J, Martin M and Maier J 2005 *Solid State Ionics* **176** 1465–71

[8] Crank J 1975 *The Mathematics of Diffusion* 2nd edn (Oxford: Oxford University Press)

[9] MATLAB *R13* (Natick, MA: The Mathworks Inc)

[10] DeSouza R and Martin M 2009 *MRS Bull.* **34** 907–14

[11] Henderson A 2013 Multivariate analysis of SIMS spectra *ToF-SIMS Material Analysis by Mass Spectrom.* 2nd edn ed J C Vickerman and D Briggs (IM Publications and Surface Spectra) pp 449–83

[12] Heller-Krippendorf D 2019 *Mulitivariate Data Analysis for Root Cause Analyses and Time-of-Flight Secondary Ion Mass Spectrometry* (Wiesbaden: Springer Spektrum)

[13] Graham D, Gamble J and L J 2023 *Bioninterhphases* **18** 031201

[14] Hook A L, Chien C-Y, Yang J, Atkinson S, Langer R, Anderson D G, Davies M C, Williams P and Alexander M R 2013 *Adv. Mater.* **25** 2542–7

[15] Fricker A L, MacPhail D S, Keneghan B and Pretzle B 2017 *NPJ Herit. Sci.* **5** 28

Sarah Fearn

Chapter 5

Applications

Secondary ion mass spectrometry (SIMS) has many strengths as a characterisation technique, which is highlighted by the various ways SIMS has been applied to a vast array of materials. Primarily, the ability of SIMS to measure all elemental species and their isotopes, as well as molecular species, makes it unique among characterisation techniques. Being able to maintain the sample in its original solid state allows compositional analyses to be obtained over all three dimensions, i.e. the *x*-, *y*-, and *z*-planes. Coupled with this, the ability to analyse both conducting and insulating materials broadens the technique's appeal further. SIMS can therefore provide an overwhelming amount of data over varying length scales and dimensions.

Typically, when the applications of SIMS are discussed, they are presented in clearly defined material categories, such as semiconductors, ceramics, metals, and glasses. However, over the last ten years or so, the areas of application have not been so easily and readily defined. As SIMS technology has changed and developed, coinciding with many new developments in materials science such as polymer electronics, biomaterials, and hybrid battery materials, its possible applications have also expanded enormously, which is a testament to the underlying flexibility of the technique. More commonly, SIMS is not used as a standalone analytical technique to find a solution but is rather used as part of a multimodal approach to characterising materials, particularly in the field of biological and medical studies.

It would be difficult to review all the applications of SIMS in materials science, as they are indeed numerous. In the following sections, the application of SIMS to a collection of different materials has been highlighted, and the selection is in no way comprehensive. There is a large body of work in conservation science, where SIMS has been used to analyse painting samples to understand the chemical makeup of paints used by artists for a greater understanding of artistic practice. Degradation and corrosion studies of objects made from metals, glass, and early plastics help in the preservation of rare and precious objects [1–5]. Another field where SIMS is

doi:10.1088/978-0-7503-3331-3ch5

routinely applied is in the earth sciences, where the very high mass resolving power of large magnetic sector instruments is applied to great effect in studies ranging from geochronology, trace element, and stable isotope analyses [6, 7]. Cosmochemistry has benefited from the small spot size of NanoSIMS and focused ion beam SIMS (FIB-SIMS) instruments to investigate very small particles in space dust and meteorites [8, 9]. SIMS has also been used to investigate wood composition [10]. Finally, SIMS is a developing application in forensic science [11].

5.1 Functional devices

For several decades, the analysis of electronic devices, mainly those fabricated from semiconductor materials, was very much the 'bread and butter' of dynamic SIMS analyses. The ability to collect SIMS depth profiles with very high depth resolution from the top 3 nm up to several hundred nanometres, coupled with high chemical sensitivity, helped to drive the miniaturisation of many electronic devices and push the limits of the microelectronics industry. Early work was very strongly focussed on the quantification of implant depth profiles, SiOx layers, and interfaces and has been thoroughly written about in the literature [12]. More recently, with the advent of quantum computing, ReRAM devices, and perovskite-based devices, SIMS has again become an important tool for the advancement of these materials.

5.1.1 Implant profiles for quantum applications

For years, SIMS has been extensively used within the semiconductor industry to measure and quantify implants, profiles, and layer interfaces, with SIMS instruments being a fundamental part of many in-line semiconductor fabrication facilities. The unparalleled detection limits and large dynamic ranges achievable in small analytical areas have made SIMS the cornerstone technique for the electronics industry. It is essential to be able to obtain accurate and precise dopant profile measurements as quickly as possible, as variations in dopant profiles can lead to poor device operation. Along with extensive use in industrial fabrication facilities, SIMS is still used in the research and development of semiconductor materials.

In recent years, the growth and possibilities of quantum computing have necessitated the development of novel devices [13] and new growth protocols to produce ever shallower and more tightly confined implant profiles on the atomic scale. Controlled atomic-scale patterning of phosphorous dopant placement on silicon has previously been demonstrated using phosphine as the precursor gas [14]. More recently, this has been achieved using a scanning tunnelling microscopy (STM) tip to pattern a monolayer hydrogen mask by placing arsenic atoms on a Si (001) surface using arsenine as the precursor [15]. The final step in device fabrication is to encapsulate the dopant layer in crystalline Si grown via molecular beam epitaxy (MBE). SIMS depth profiling has been used to characterise an arsenic δ-layer after Si overgrowth to identify the best fabrication conditions (see figure 5.1). In figures 5.1 (a)–(e), it can be seen that the arsenic δ-layer segregates, except for figure 5.1(c), where good confinement is achieved but the temperature is too low to form crystalline Si.

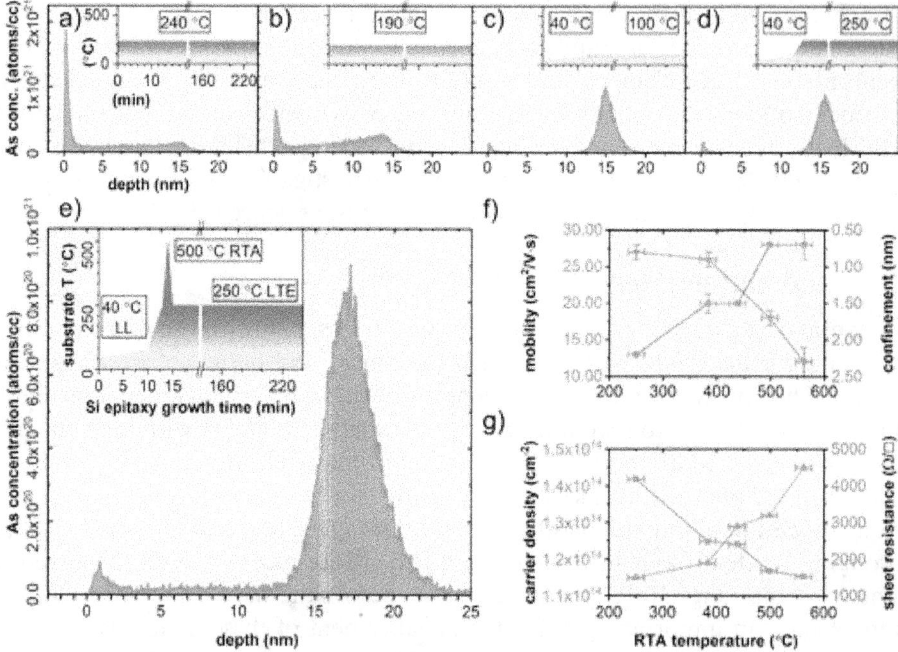

Figure 5.1. Arsenic δ-layer distribution and electronic transport controlled through a silicon overgrowth thermal annealing programme. (a)–(e) SIMS depth profiles of five arsenic δ-layer samples with different saturations overgrown with 15 nm of Si at a deposition rate of 1 monolayer per minute. (f) δ-layer mobility and confinement as a function of temperature; (g) δ-layer carrier density and sheet resistance as a function of temperature. Reprinted with permission from [15]. Copyright (2020) American Chemical Society.

To suppress arsenic segregation, a silicon 'locking layer' was introduced (see figure 5.1(e)), whereby a very thin reduced-temperature layer was grown in the first instance; then, the temperature was rapidly increased to 250 °C for the rest of the Si overgrowth (see the inset of figure 5.1(e)). The SIMS profile clearly shows the confinement of the arsenic δ-layer, successfully demonstrating that *n*-type dopants based on arsenic can be fabricated using STM H-lithography.

5.1.2 Redox resistive access memory

Redox-based resistive random-access memory (ReRAM) devices have recently been gaining attention as next-generation nonvolatile memories and important memristive components for many developing computing systems. ReRAMs are composed of a very thin silicon oxide (SiO_x) layer (between 5 and 30 nm) sandwiched between two electrodes (typically Au, Ti, or Mo). As the oxide layer is sub-stoichiometric, reversible resistance switching occurs via soft dielectric breakdown. However, device reliability is a serious concern, as identically produced devices have been observed to have varying performances, with the presence of hydrogen suspected as the issue, based on previous studies carried out in the gate oxide community [16–18]. Despite significant concerns about the presence and role of hydrogen in both ReRAMs and gate oxides, it is often neglected due to the significant challenge of measuring hydrogen concentrations. Another obstacle to device development has been measuring the small fluctuations in oxygen stoichiometry that cause changes in conductivity. In both cases, the unique strengths of SIMS, i.e. being able to measure hydrogen and being able to measure chemical changes over the nanometre range, make it a useful technique with which to interrogate ReRAM materials.

To perform precise SIMS measurements to monitor the oxygen changes occurring in a ReRAM device after biasing, a careful assessment of the experimental conditions was made [19]. Along with ensuring that appropriate ion beam conditions were selected that imparted the minimum amount of ion beam damage and the greatest secondary ion yield, other factors such as ion beam current, focus, and main chamber pressure were tightly controlled, giving a precision between measurements of ~10% [20]. To further reduce the error to below the level of the changes that needed to be measured, it was found that all the depth profile measurements needed to be done at the same time. A schematic of the sample setup is shown in figure 5.2 (a). The black outline shows the 40 μm^2 SIMS analytical area used for depth profiling. Included in the area are two pristine strips used to calibrate the data for *in situ* normalisation, and two strips that were electrically biased using a conducting atomic force microscope (CAFM). The SIMS data from the biased strips was point-to-point normalised to the data from the pristine strips, minimising matrix effects and variations between measurements. Different metal electrodes were used beneath the SiO_x layer, namely Ti, Mo, and Pt.

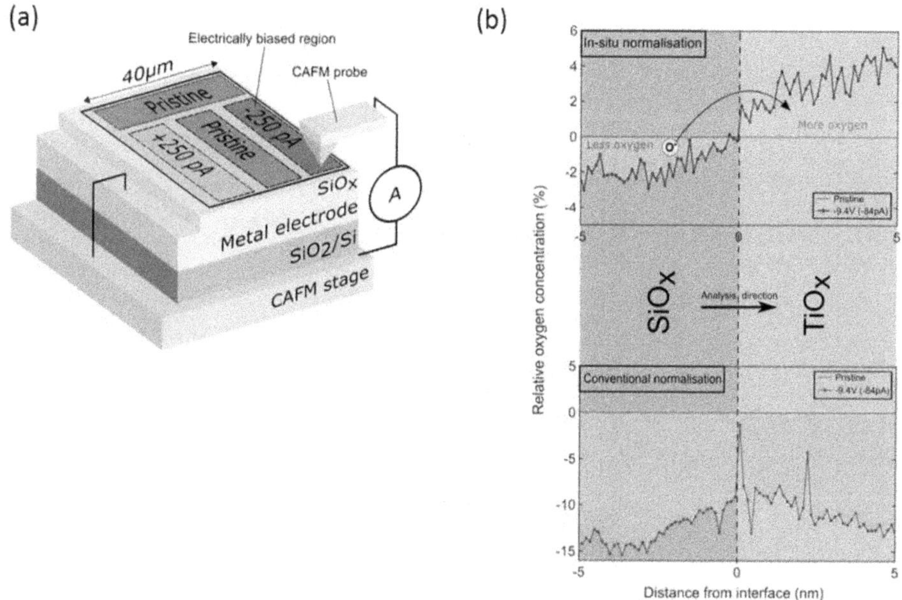

Figure 5.2. Schematic of the experimental area used for SIMS depth profiling to obtain compositional changes across the SiO_x–metal electrode interface after biasing two regions in a CAFM. Reprinted from [19], with the permission of AIP Publishing.

The improved depth profile obtained using the *in situ* normalisation is represented by the blue line in figure 5.2(b). The change in oxygen concentration across the interface is clearly revealed, compared to the profile produced by conventional normalisation (purple line). The errors in the conventional normalisation process falsely indicate that biasing reduces the oxygen concentration across the interface. Using these optimised analytical conditions, a range of samples was then analysed using a 1 keV Cs^+ ion beam at 70 nA at 45° to the samples' normal angles. Secondary ions were generated by a 25 keV Bi^+ ion beam at 45°, and charge compensation was provided by a low-energy electron flood gun. *In situ* normalised oxygen depth profiles through negatively and positively biased SiO_x–Mo interfaces (labelled MoO_x due to the large amount of oxygen) are shown in figure 5.3. The curved arrows on the plots indicate the oxygen diffusion due to the CAFM biasing step.

The development of the careful SIMS analysis revealed features that had not been previously observed. The profiles showed that the Mo electrode displayed reservoir-like behaviour in exchanging oxygen with the SiO_x layer under opposite biases, and crucially, it prevented device failure and delamination [19]. Following on from

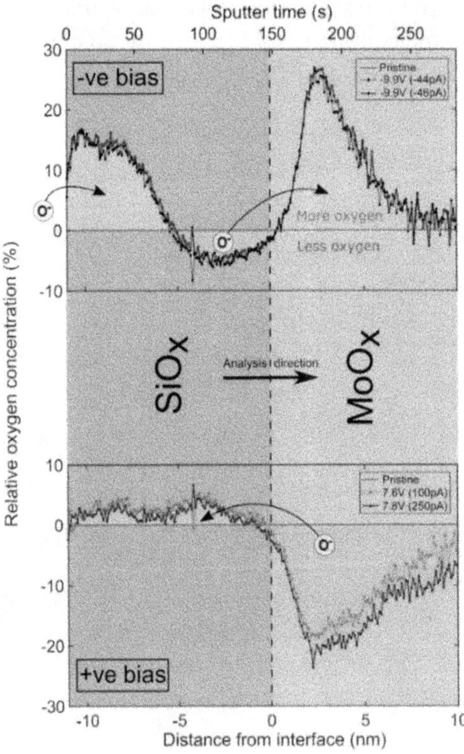

Figure 5.3. In-situ normalised oxygen depth profiles through negatively (top) and positively (bottom) biased SiO_x–MoO_x interfaces. Curved arrows show the direction of oxygen diffusion due to the biasing. Reprinted from [19], with the permission of AIP Publishing.

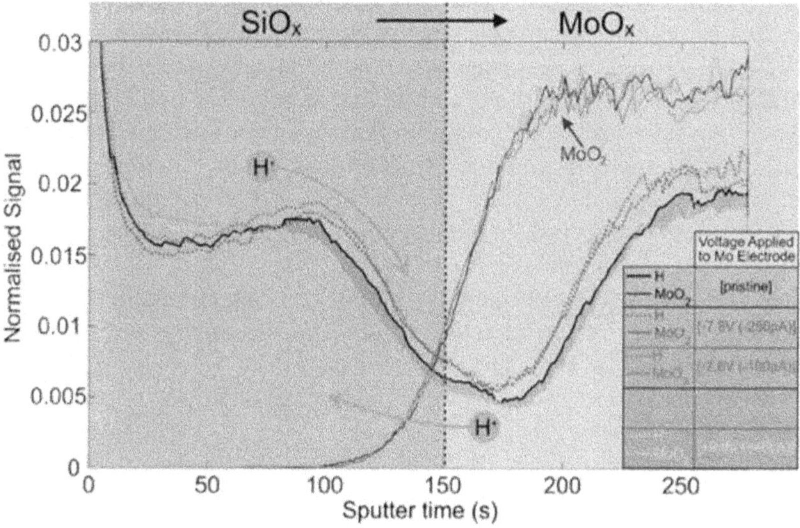

Figure 5.4. SIMS depth profiles of hydrogen across a SiO_x–MoO_x interface; a pristine sample and samples under negative and positive biasing of the Mo electrode [21] John Wiley & Sons. [© 2024 The Author(s). Advanced Materials published by Wiley-VCH GmbH].

measuring the oxygen transport in the SiO_x–Mo system, the same methodology was applied to measure the evolution and movement of hydrogen across the same interface after biasing to simulate ReRAM operation [21]. The same SIMS analytical conditions were applied as before.

The SIMS depth profiles for hydrogen across pristine and biased SiOx–MoOx interfaces are shown in figure 5.4. The green highlighted region indicates an increase in hydrogen content when the Mo electrode is under negative bias, whereas the red highlighted region indicates a reduction in hydrogen content when the electrode is positively biased. For the first time, this work confirms the reversible diffusion of hydrogen across an interface after ReRAM operation.

The authors also compared the SIMS depth profiles with IV sweeps across the same samples. The electroforming phase occurred with a positive bias on the Mo electrode, which indicates a reduction in hydrogen content. During the reset, when the bias is reversed (negative) and conductivity reduced, an increase in hydrogen content is observed in the SIMS profile, which may indicate that hydrogen is possibly passivating defects at the interface [10].

In conclusion, the careful SIMS analysis and experimental setup were able to reveal previously unknown mechanisms related to the roles of both oxygen and hydrogen species in the operation of ReRAM devices.

5.2 Metals and alloys

SIMS has a long history of being applied to metals and alloys, from the very early work looking at how metal ions are formed under ion beam bombardment to the ion mapping of complex alloys [22, 23]. SIMS has been able to provide useful insights into the causes and mechanisms of many environmental degradation processes.

The material properties of many advanced metal alloys are dependent on low concentrations of certain elements and/or the presence of light elements such as hydrogen. Other issues related to material failure, such as fatigue cracking, require chemical information to be obtained from cracks, grain boundaries, or localised precipitates. In all these cases, the ability of SIMS to detect light elements, its high sensitivity, and highly resolved chemico-spatial capabilities lend themselves perfectly to analysing this group of materials.

The samples themselves also have the benefit of being conductive, so there is no deterioration of data due to charging effects, and in many cases, very smooth polished surfaces can be prepared or ion-beam milled to select specific areas of interest for analysis. Some of the unique applications of SIMS to metals and alloys are highlighted here.

5.2.1 Hydrogen embrittlement

Titanium alloys are widely used in extreme engineering applications due to their excellent properties of corrosion resistance, high specific strength, and fatigue limit strength-to-weight ratio [24]. They have been central to advances in both jet engines and the aerospace industry. Hydrogen embrittlement (HE) is, however, a major issue in their continued development, as it can cause catastrophic failure of essential components. Despite the presence of a native oxide film on the alloy, hydrogen is still observed in the alloy microstructure. One way to reduce the probability of HE is to reduce (ideally stop) the hydrogen entering the alloy structure, and to this end, it is essential to understand the mechanism behind the adsorption/absorption of hydrogen at the surface.

Time-of-flight SIMS (ToF-SIMS) was carried out on the surfaces of a Ti–6Al–4V alloy that had been polished and roughened and then exposed to room-temperature electrochemical H-charging [25]. Microscopic analyses and thermal desorption spectroscopy revealed that the surface roughening increased the recombination of atomic hydrogen to molecular hydrogen, thus reducing atomic hydrogen uptake into the alloy [26]. The ToF-SIMS data further revealed that the high defect density underneath the roughened surface impedes further hydrogen ingress into the bulk (see figure 5.5).

Other metal alloys such as steel are also vulnerable to HE, but again, elucidating the role hydrogen plays in this process has proved difficult due to the small analytical areas and volumes involved. Identifying hydrogen at potential hydrogen trapping sites such as grain boundaries, crack tips, and dislocations is not straightforward. By carrying out fatigue tests using atmospheres rich in the isotope deuterium (^{2}H), it becomes possible to monitor the distribution of deuterium in the steel as an unambiguous marker for how hydrogen would behave under the same test conditions. High-resolution NanoSIMS imaging can then be used to map the deuterium distribution at the tip of and in the wake of secondary and tertiary fatigue cracks [27, 28]. The distributions of deuterium and oxygen are shown in

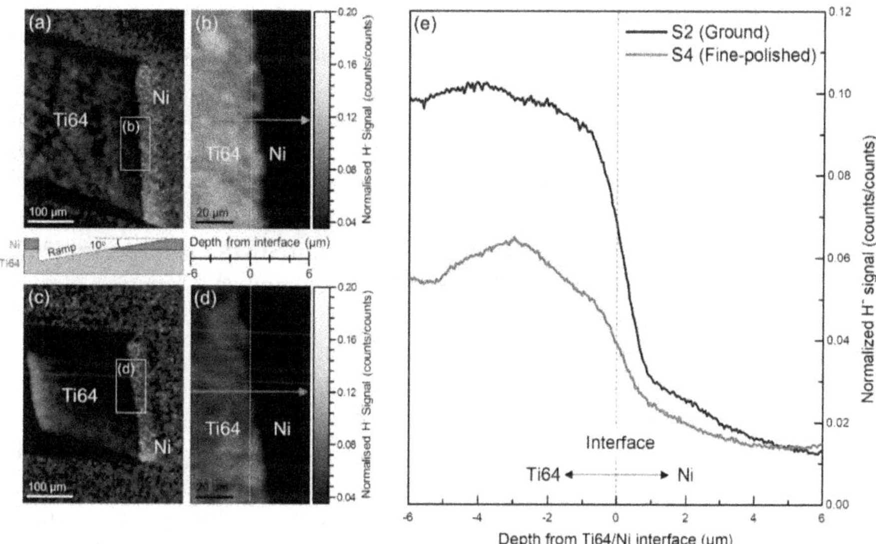

Figure 5.5. SIMS ion maps for (a, b) polished samples and the corresponding SIMS H ion map and (c, d) unpolished samples and the corresponding SIMS H ion map (Ni is present as a protective layer on the sample surface). (e) Hydrogen profile obtained from H ion maps. Reprinted with permission from [26], Copyright (2021), with permission from Elsevier.

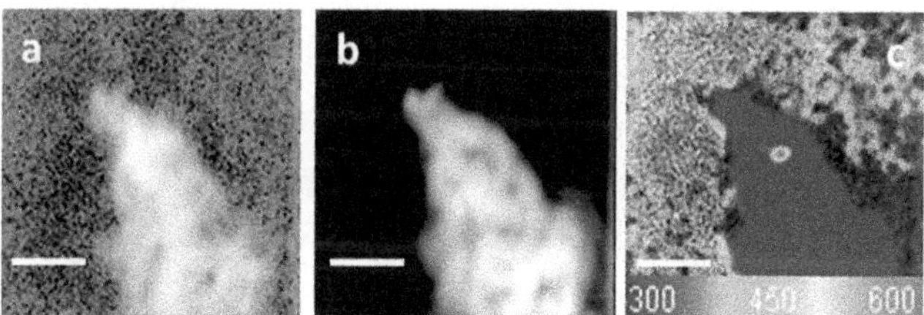

Figure 5.6. High-resolution NanoSIMS isotopic imaging of 316LN steel: (a) ^{2}H and (b) ^{16}O ion images and (c) an HSI of the $^{2}H/^{16}O$ ratio. The scale bar is 1 μm. Reproduced from [27], with permission from Springer Nature.

figures 5.6(a) and (b), respectively, and the ratio of the two species is shown in the hue saturation image (HSI) of figure 5.6(c). Significantly, the HSI image highlights two red regions at the front of the crack tip and to the side and locates hydrogen in the region of the crack deformation structure, indicating how hydrogen is influencing the formation of the fatigue crack.

5.2.2 Corrosion

Very localised corrosion and pitting corrosion are among the most common and catastrophic causes of failure in metal structures. Similar to HE, it has been difficult to obtain useful chemical data associated with this localised form of corrosion due to the small areas available and the problem of finding the specific areas to analyse. The application of liquid metal ion guns (LMIGs) in SIMS instrumentation has played a pivotal role in enabling analysis-site-specific SIMS. The highly focussed, high-brightness beams are able to remove material to uncover features of interest and then probe the sample to capture secondary ions.

In stainless steel materials, the majority of pitting corrosion events are found to occur at or around second-phase particles formed during manufacture. Nanoscale SIMS analysis around MnS particles in a 316 F stainless steel was performed using an FEI FIB 200 with a SIMS attachment. The highly focussed Ga ion beam (100 nm spot size) was used as either a spot or annulus to obtain SIMS mass spectra [29]. The local chemical changes (variation in Cr:Fe ratio) occurring as a function of distance from the edge of an inclusion, as well as chemical variation within the inclusions themselves were observed. A localised zone of chemical variation of between 200 and 400 nm is clearly shown, where the Cr level over this region is depleted to levels low enough to render the steel susceptible to corrosion pitting.

The ability to carry out site-specific SIMS analysis with a focused ion beam (FIB)-based SIMS was also used to identify the chemical composition of an unknown blue-coloured fatigue crack initiation feature that was observed during low-temperature testing of a titanium superalloy of Ti–6Al–2Sn–4Zr–6Mo (Ti-6246) [30]. The blue spot is shown in the electron image of figure 5.7(a). FIB-SIMS analysis was carried out in the regions highlighted in figure 5.7(b). The chemical profiles in figure 5.7(c) show that in the region of the blue spot, the near surface of the fracture edge (the top 20 μm) demonstrates elevated levels of Na^+, Cl^-, and O^- ions. The identification of

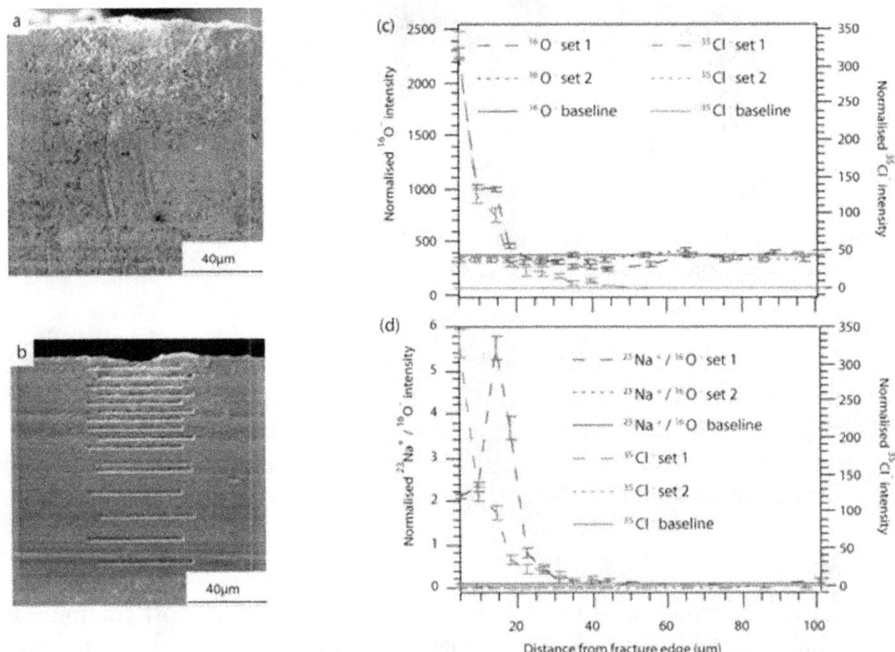

Figure 5.7. (a) Blue spot corrosion site. (b) FIB-SIMS analysis regions. (c) and (d) FIB-SIMS data obtained from the edge of the fracture surface though the blue spot into the bulk material. Reprinted with permission from [30], Copyright (2015), with permission from Elsevier.

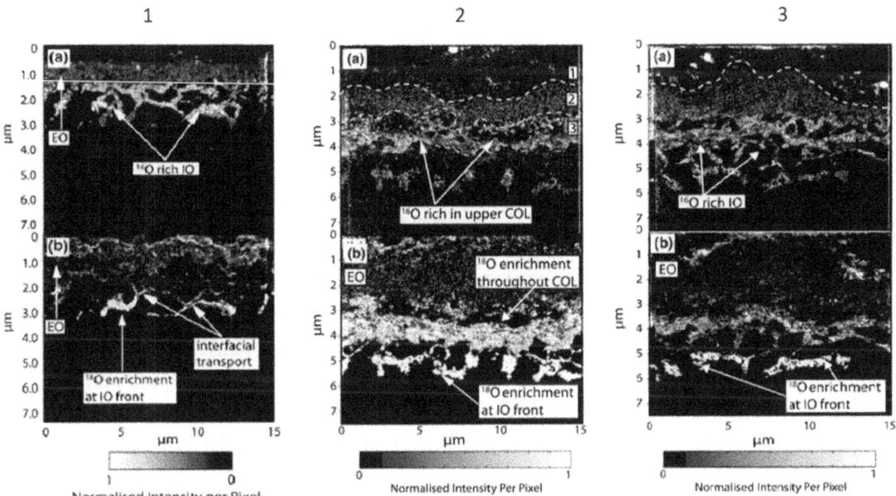

Figure 5.8. SIMS ion maps: 1. the isotope-exchanged samples (a) $^{16}O^-$ and (b) $^{18}O^-$; 2. the sample under compression (a) $^{16}O^-$ and (b) $^{18}O^-$; 3. the sample under tension (a) $^{16}O^-$ and (b) $^{18}O^-$. Reproduced from[31], with permission from Springer Nature.

these ions within the blue spot highlighted the process of hot salt corrosion cracking (HSCC) with contamination from poor handling of the sample. It was noted that although this had been a nucleation point for fatigue cracking in a lab-based environment, this is not observed for the same materials when used under the higher pressures and temperatures of real-world operating conditions.

The formation of oxide scales on metals can be considered a form of corrosion, or, depending on their structure and composition, protective layers. This is particularly true for Ni-based superalloys used in high-temperature turbine applications. Excessive oxidation of components fabricated from these alloys can reduce a component's mechanical integrity. The oxidation resistance of the alloy depends on the alloy's ability to form a compact and adherent scale. To investigate the oxidation behaviour of a fine-grained Ni-based superalloy, an ^{18}O–^{16}O isotope exchange experiment was carried out, followed by SIMS analysis of the oxide layer formed [31].

The SIMS ion maps showed that applied stresses had very little influence on the inward oxygen diffusion and demonstrated a build-up of ^{18}O at existing ^{16}O internal oxide sites due to rapid diffusion along the internal oxide/alloy interface (see figure 5.8). Previously, the oxidation behaviour of these superalloys was thought to be due to stress-assisted grain boundary diffusion; however, the SIMS analysis presented in this study clearly demonstrated a different oxidation behaviour.

5.3 Energy-related materials

The flexibility of SIMS has made it a powerful analytical tool for a range of energy-related materials. For many years, it has been used to investigate the diffusion properties of materials used in the cathodes of solid-oxide fuel cells (SOFCs). More recently, as the importance of efficient renewable energies has grown, the range of materials has expanded to include catalysts for the electrochemical conversion of biomass, perovskite solar cells, and lithium-ion- and sodium-ion-based battery materials [32–35].

5.3.1 Battery materials

Studies of lithium- and sodium-ion batteries are focused on finding the optimal cathode, anode, and electrolyte materials that produce high performance and are stable. Several issues are encountered in Li- and Na-metal cells: dendrite formation that causes short circuits, the formation of dead metal regions, continuous solid electrolyte interphase (SEI) formation that consumes the electrolyte, and gas evolution, to name a few [36]. A large body of work is available on the application of SIMS to battery materials, encompassing solid-state lithium [37], electrolyte interfaces in lithium-ion batteries [38], and overviews of best practices for applying SIMS to battery materials [39, 40].

Among the issues that affect battery performance the most, the formation of the SEI has been studied extensively [41–44]. The formation of the SEI layer is essential for successful Li or Na batteries. Upon contact between the metal and the electrolyte solution, a layer forms instantaneously, which consists of soluble and insoluble reduction products of the electrolyte [45]. The SEI is a key factor that determines the safety, power capability, shelf life, and cycle life of a battery. It must also be both mechanically stable and flexible. Therefore, understanding the formation and deterioration of the SEI is fundamental to improving the safety and lifetime of Li and Na batteries. Analysing this region, however, is not trivial: samples are typically air-sensitive, so sample preparation can be difficult; the analytical regions of interest are very small, and the data can be complex.

Figure 5.9 shows the mass spectra of a Mo electrode after electrochemical treatment in a $LiClO_4$ solution [46]. It is clear that the mass difference between each highlighted group is ~16 amu; thus, the difference between the groups is most likely to be due to the oxygen in the complex. It is known that Cl has two stable isotopes (^{35}Cl and ^{37}Cl) and Mo has seven stable isotopes (^{92}Mo, ^{94}Mo, ^{95}Mo, ^{96}Mo, ^{97}Mo, ^{98}Mo, and ^{100}Mo); furthermore, in SIMS, the peak intensity of isotopes corresponds to the abundance of each isotope. Thus, the peak groups in figure 5.9(a) are identified as Cl^-, ClO^-, ClO_2^-, ClO_3^-, and ClO_4^-, and the peak groups in figure 5.9 (b) are identified as MoO^-, MoO_2^-, MoO_3^-, and MoO_4^-. In this scenario, the isotope patterns of both Mo and Cl help to identify the peaks in the mass spectra, and there are not too many mass interferences.

However, peak interferences can mislead the data interpretation, and extra care should be taken during the analysis of mass spectra data, not only at higher *m*/*z* ratios. For example, in the analysis of $Li(Ni,Mn,Co)O_2$ (NMC) cathodes from Li-ion batteries made with a polyvinylidene fluoride (PVDF) binder and conductive carbon, the masses of the LiF^- and $C_2H_2^-$ species are 26.0178 and 26.0156 amu, respectively. In order to separate these, a mass resolution of more than 11 826 is required. This is beyond the capability of most ToF-SIMS instruments, as

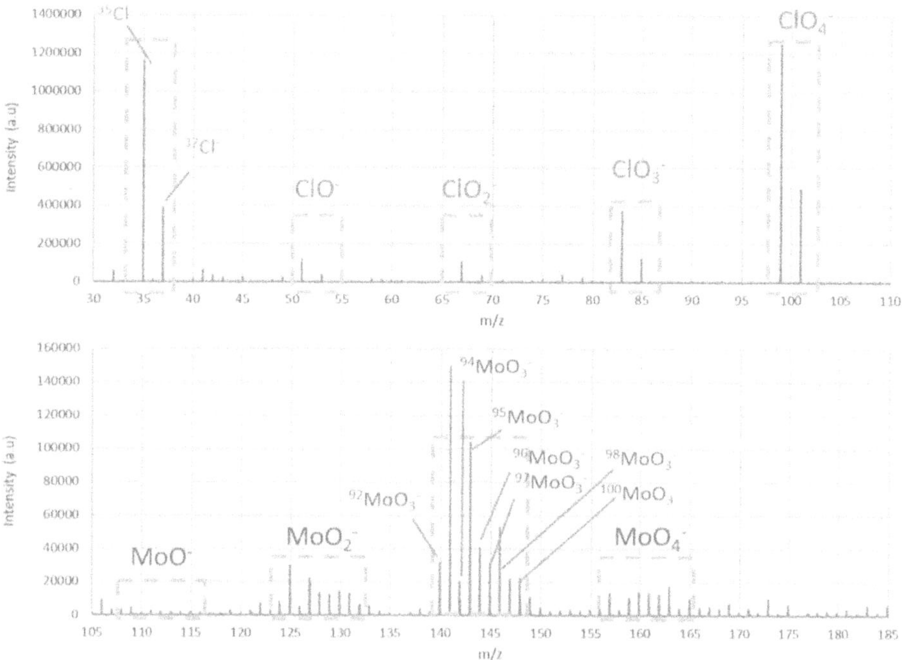

Figure 5.9. Selected regions of SIMS mass spectra of Mo electrodes after an electrochemical reaction in $LiClO_4$ solution. The spectra were obtained using an Ar_n^+ sputtering beam in the negative mode with: (a) a mass range of 30–110 amu and (b) a mass range of 105–185 amu. Reproduced from [39], with permission from Springer Nature.

topographical issues such as roughness will undoubtedly result in broadened peaks, further reducing the resolution of the data. In this situation, it may be possible to use species that originate from the same component or are correlated. Isotopes can be potentially chosen as alternative options. In this case, one could use C_2H^- or $C_2H_3^-$ instead of $C_2H_2^-$; similarly, $^6LiF^-$, $^7LiF_2^-$, or $^7Li_2F_3^-$ could be selected to replace $^7LiF^-$.

When investigating the depth distributions of organic/inorganic molecular species, clarification of peak assignment can also be helped by analysing reference compounds beforehand to obtain a clearer picture of the main peaks of interest that should observed [47]. Table 5.1 shows a list of important secondary ion peaks obtained from the following reference materials: LiTiO powder, PVdF binder, carbon additives (CB), $LiPF_6$ electrolyte salt, LiF, and Li_2CO_3.

These peaks were then used to help identify the depth distributions of organic and inorganic species forming in the SEI of $Li_4Ti_5O_{12}/LiNi_{3/5}Co_{1/5}Mn_{1/5}O_2$ (LTO/NMC) and $Li_4Ti_5O_{12}/LiMn_2O_4$ (LTO/LMO) after galvanostatic cycling. The signals were then plotted as 3D depth profiles (see figure 5.10). The images clearly show a much thicker SEI layer on the LTO anode facing the LMO electrode compared to the SEI layer on the LTO anode facing the NMC electrode. The thicker SEI layer would impede the Li diffusion in the system and would explain the decreased capacity of the LTO/LMO system.

Table 5.1. Main secondary positive ion peaks detected in ToF-SIMS from reference powders of LiTiO, PVdF, CB, $LiPF_6$, LiF, and Li_2CO_3. Reprinted from [47], Copyright (2020), with permission from Elsevier.

LTO powder		Reference PVdF		Reference CB		References $LiPF_6$/LiF		Reference Li_2CO_3	
m/*z*	ID	*m*/*z*	ID	*m*/*z*	ID	*m*/*z*	ID	*m*/*z*	ID
7	Li	12	C	13	CH	14	Li_2		
27	C_2H_3	31	CF	14	CH_2	19	F		
30	Li_2O	41	H_3F_2	15	CH_3			30	Li_2O
39	K	43	C_3H_3O	26	C_2H_2	33	Li_2F	31	Li_2OH
48	Ti	51	CHF_2	27	C_2H_3	59	Li_3F_2	37	Li_3O
64	TiO	69	CF_3	29					
97	TiO_3H	77	$C_3H_3F_2$		C_2H_5	85	Li_4F_3	81	Li_3CO_3

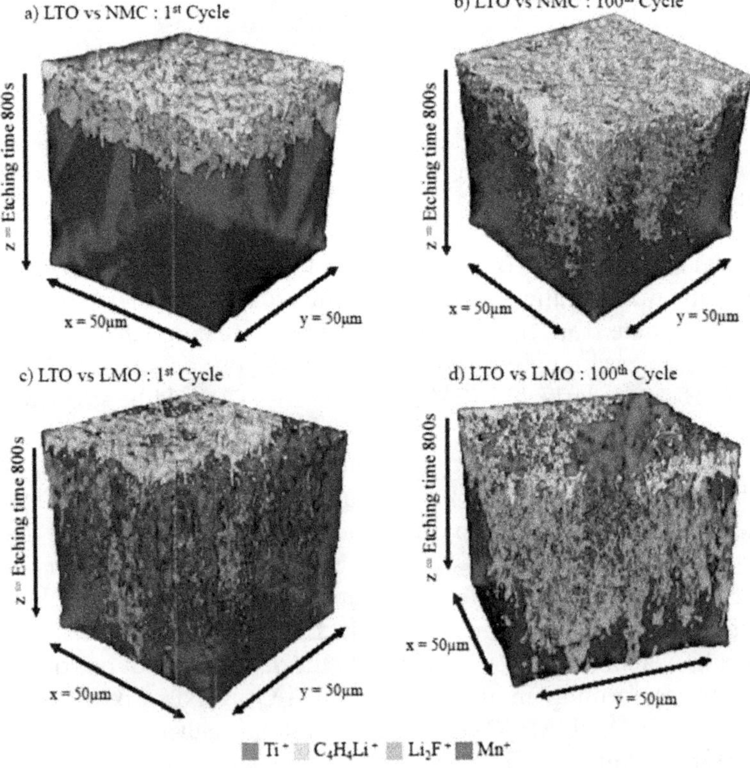

Figure 5.10. 3D ToF-SIMS depth profiles showing the spatial distribution of Ti^+, $C_4H_4Li^+$, Li_2F^+, and Mn^+ secondary ions at the top of the LTO electrode at the 1st and 100th cycle: (a) and (b) LTO/NMC and (c) and (d) LTO/LMO. Reprinted from [47], Copyright (2020), with permission from Elsevier.

The 3D ion images in figure 5.10 nicely show the power of SIMS in mapping the spatial distributions of ions of interest and revealing fundamental materials processes. As mentioned previously, interrogating SEI regions and Li- and Na-based batteries more generally is made harder due to the small regions of analytical interest and their air sensitivity. To overcome these drawbacks, it is possible to use a plasma focussed ion beam (PFIB) to mill into a sample, expose a region of interest, and then subsequently use SIMS to make an ion map of the freshly exposed area. Revealing an area under vacuum using an ion beam allows the sample to be air-handled.

Focussed ion beam (FIB) milling was carried out on a polycrystalline Li $(Ni_{0.8}Co_{0.1}Mn_{0.1})O_2$ (NMC811) cathode material that had been cycled. The sample was mounted on a pre-tilted sample holder in the Hi5 PFIB SIMS instrument (Imperial College, London) [48]. The sample stage was also tilted relative to the ion beam (figure 5.11(a)) to create an angled trench in the sample and to minimise ion

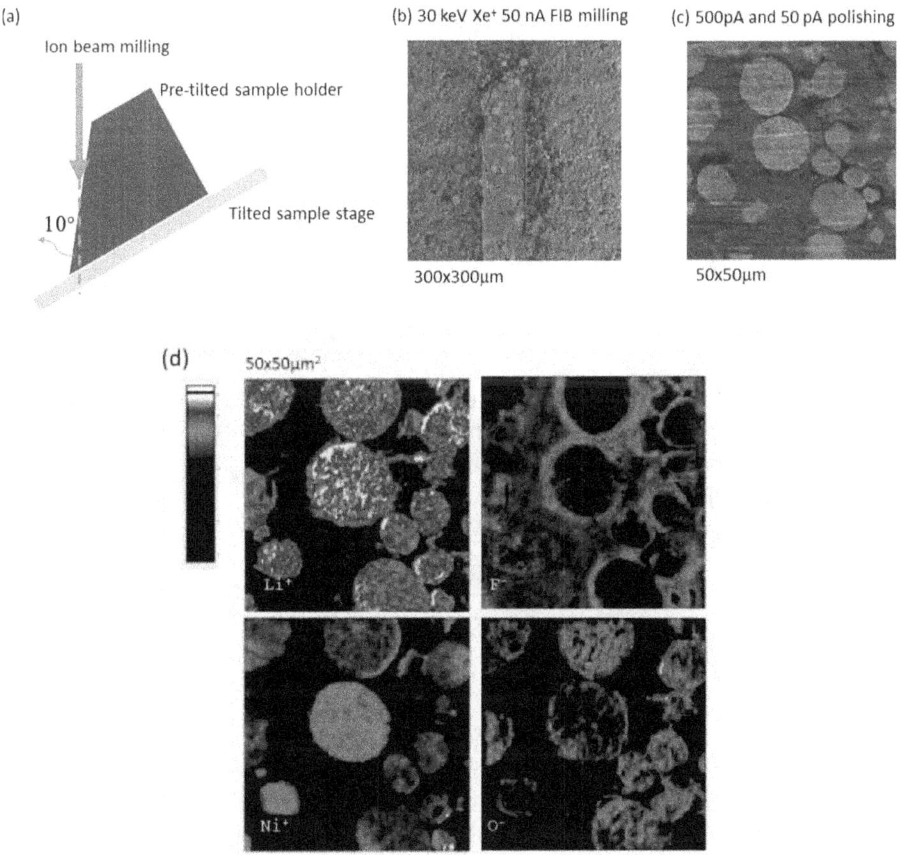

Figure 5.11. Ion beam milling and SIMS analysis of NMC811: (a) PFIB milling setup; SEM images of (b) ion beam-milled surface and (c) after ion beam polishing; (d) SIMS ion images of O^-, F^-, Li^+, and Ni^+ from the milled surface, obtained via simultaneous positive and negative secondary ion collection (data courtesy of Dr Zonghao Shen).

beam damage to the milled surface. The trench was milled along the sample using a 30 keV Xe^+ PFIB (see figure 5.1(b)) to expose a fresh surface. The surface was then further ion beam polished using a series of reduced ion beam currents to achieve a smooth surface (see figure 5.11(c)). Once the surface was prepared, SIMS ion images were taken over the milled surface. Both positive and secondary ions were collected simultaneously, minimising the area needed for data collection [49]. The ion images are shown in figure 5.11(d).

The ion images clearly show the distribution of O^-, F^-, Li^+, and Ni^+ throughout the section material. The Li^+ and Ni^+ ions are colocated in the main body of the spheres, and F^- is observed decorating the outer edges of the spheres. Interestingly, O^- is not homogenous in the ion map, and in some cases, it is only visible around the edges of some spheres and appears depleted.

The advantage of this procedure is that a fresh surface is fabricated that has not been exposed to air or any contamination from a glove box during sample mounting. The highly focussed plasma ion beam is also able to produce highly resolved images due to its very small spot size (see section 2.1.1.3).

The highlighted studies and the available literature collectively demonstrate the versatility and power of SIMS in probing the complex chemical and structural aspects of SEI formation in various battery systems. SIMS can provide detailed insights into the composition and evolution of the SEI that is crucial for the development of more efficient and durable energy storage systems.

5.3.2 Nuclear waste glass

Still on the horizon as a useful alternative to fossil fuel, nuclear energy supplies approximately 14% of the UK's total electricity supply. The UK currently has a total of ~5 million m^3 of nuclear waste, of which 3% is made up of high-level waste (HLW), characterised by high levels of radioactivity and decay heat. A future strategy for the immobilisation of nuclear HLW is to vitrify it and dispose of it below ground at depths of up to 1000 m.

Dissolution studies of nuclear HLW glass are, therefore, of great interest due to the potential long-term effects of HLW on the environment. Glass instability and leaching may arise, for example, due to interactions with underground water [50]. Previously, SIMS has been used in several studies to examine the elemental distribution in corroded HLW glasses SON68 and SM513 after long-term leaching experiments [51, 52].

Heavily corroded HLW glasses, however, are not straightforward samples to depth profile. Difficulties arise for several reasons: the surfaces of the corroded samples become extremely rough due to the exchange of material during the leaching experiments, so the depth resolution of the depth profiles is very poor. Charge compensation is also necessary, as the samples are insulating. As the samples are highly insulating, the positive charge from the sputter ion gun builds up on the surface, and in some cases, the flood gun is not capable of charge compensating the samples sufficiently to enable the sputtering and analysis of the sample. The secondary ion signal drops off very quickly, and no more ions are sputtered and detected. A novel way of analysing the sample is needed, as conventional depth profiling methodologies do not always succeed.

Depth profiling of glass has been carried out for many years. It has been used to measure changing compositions to identify problems due to manufacture, the effects of processing, and storage. SIMS depth profiling has also been applied to many investigations of glass corrosion, for example, in the study of museum glass [53, 54] and in the field of nuclear waste management [55, 56].

One way to overcome the difficulties of the surface roughness of the heavily corroded samples and poor to no charge compensation is to remove the need to depth profile the samples. Similarly to the process outlined in section 5.3.1, this can be achieved by creating a bevel or slope in the sample. The composition and buried features of the sample are then exposed as magnified surface features. Bevels or slopes can be successfully created on the glass samples using an FIB or PFIB, and the resulting ion beam milled surface is very smooth compared to the top of the corroded glass surface. Once the bevel has been successfully fabricated, the sample is transferred to the SIMS instrument for ion mapping if the FIB or PFIB are not mounted within the same instrumental setup. The bevel is located by identifying the fiducial marks made during the ion beam milling process. Another advantage of ion mapping the bevelled surface is that a much lower primary beam current can now be used to ion map the surface, which again improves the charge compensation. An example of the elemental ion maps obtained from bevels made in an HLW glass which had been leached for seven days is shown in figure 5.12 [57]. The roughness and cracking of the corroded surface of the samples can be seen very clearly in the Na^+ and Li^+ ion images. The spatial distributions of the different elements within the glass are clearly visible along the region of the bevelled sample.

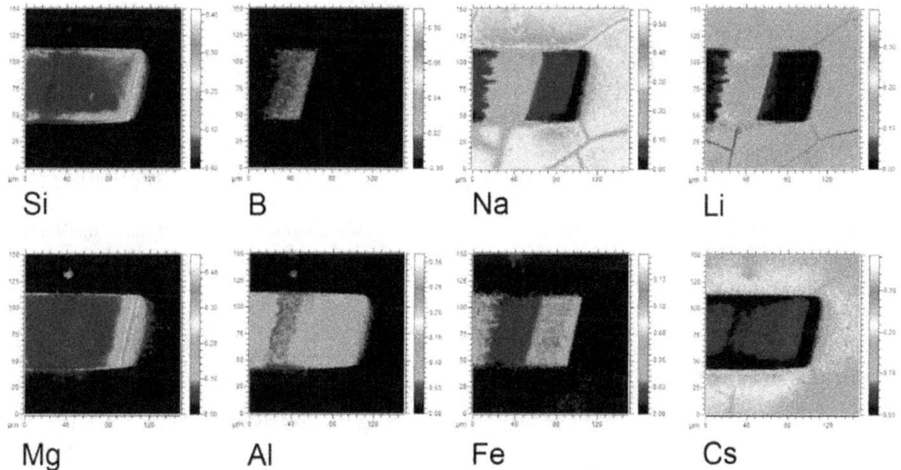

Figure 5.12. Normalised positive ion images of glass leached for seven days. Reprinted from [57], Copyright (2014), with permission from Elsevier.

As each of the individual ion images is made up of a 256 × 256 matrix of pixels (see figure 5.13(a)), each pixel corresponds to a specific ion count for the element selected (see figure 5.13(b)). This matrix of ion count data can be integrated over its width to form a single column of ion count data. By repeating the integration step for the entire set of ion images taken, a depth profile can be plotted once a depth scale has been calibrated.

To convert the length imaged along the bevel, x, shown by the green line in figure 5.14, into the corresponding depth into the material, z, the magnification of the bevel, M, must be known. This is obtained by measuring the bevel of the sample using either a surface profilometer or an interference microscope. Once the angle, θ, is measured (see figure 5.14), M can be calculated, as $M \approx \frac{1}{\theta}$ for very small angles [58].

Using the relationship in equation 5.1, where x is the length of the ion image made in the ToF-SIMS (with a certain field of view (FoV)) and θ is the angle of the bevel measured using an interferometer or stylus profilometer:

$$\text{magnification of bevel, } M = \frac{\text{distance along bevel, } x}{\text{depth into material, } z}. \tag{5.1}$$

Therefore,

$$\text{depth into material, } z = \frac{\text{distance along bevel, } x}{\text{magnification of bevel, } M}. \tag{5.2}$$

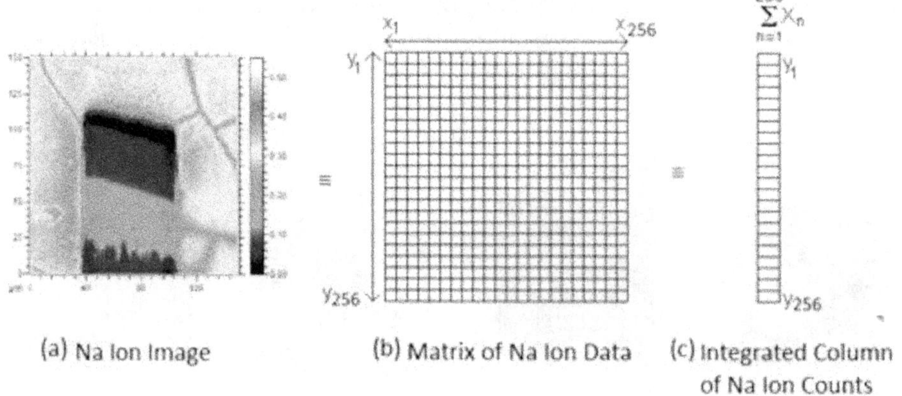

Figure 5.13. Data conversion from ion map to line scan. (a) The individual ion images are made up of a matrix of data (b) consisting of 256 × 256 pixels. This matrix may be integrated over its width (c) to form a single column of data.

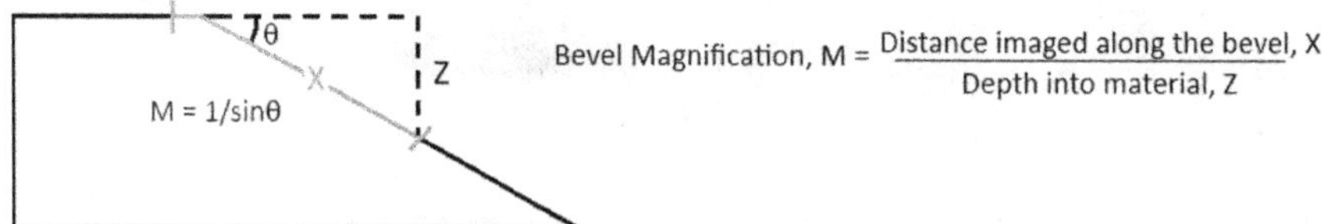

Figure 5.14. (a) The relationship between the bevel magnification, M, and bevel angle, θ, needed to convert the ion image distance x along the bevel to depth z.

Once the FoV of the SIMS ion image has been converted into the corresponding depth scale, a depth profile can be plotted, as shown in figure 5.15. The FoV on the x-axis has been converted into depth, and the different regions of interest are also marked in the figure to more clearly show the changing chemical composition in the glass due to the leaching and diffusion of the elements in the glass. A further advantage of this technique is that a much greater depth in the sample can be reached. The final depth scale of the depth profile shown in figure 5.10 is 5.70 μm. It would be impossible to depth profile up to such a depth using SIMS analysis, as the depth resolution would be quickly lost within the first micron due to the initial roughness of the sample surface.

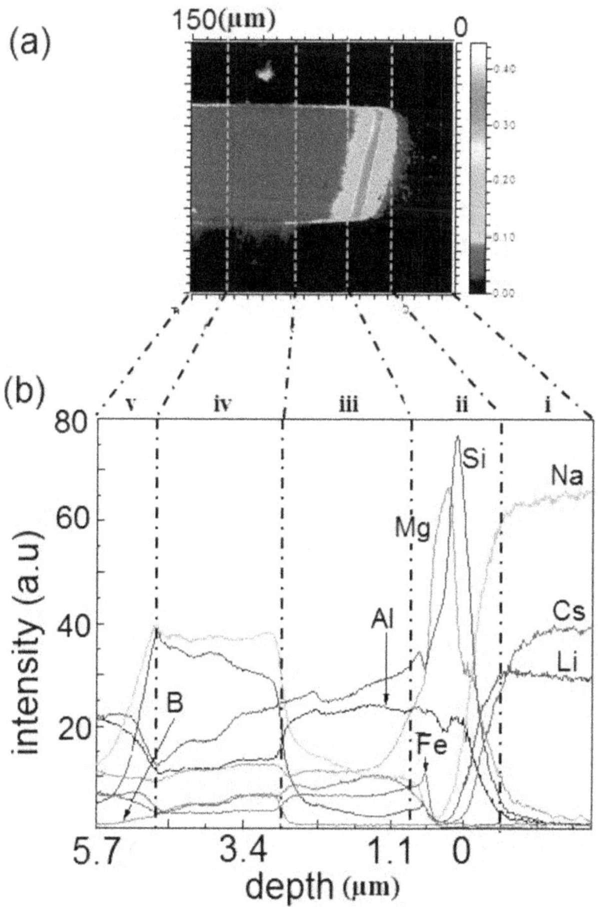

Figure 5.15. The depth profile obtained from an ion map–line scan ToF-SIMS analysis of a 5° bevel made in HLW glass via FIB milling. The length of the ion image over the bevel is converted into depth. Reprinted from [57], Copyright (2014), with permission from Elsevier.

5.3.3 Note

This method of sample preparation based on the use of an FIB to prepare analytical regions, outlined above and previously in the battery materials section, may also be useful for other samples if the initial surface is too rough and too insulating. It can also be applied to biological material to remove or slice cells to expose cross sections [59].

5.4 Biosciences

Perhaps not commonly associated with the biosciences compared to other techniques often used in this area, such as spectroscopic or optical methods, SIMS has several advantages: it can identify a wide range of analytes (drugs, proteins, metabolites, and trace elements), which is extremely useful, for example, in monitoring drug uptake in diseased tissues or specific cells such as cancer tumours. In contrast to many other techniques, SIMS does not require the use of chemical tags or fluorescent markers to identify multiple metabolites. This removes the issue of interference between labels. Both molecular and elemental ions are collected during a SIMS analysis; therefore, signals originating from both soft and mineral-based tissue (e.g. bone) can be analysed at the same time. Furthermore, the high lateral resolving power of SIMS can provide subcellular imaging capabilities over many length scales, from micrometres to centimetres, as well as 3D rendered images.

Applications of SIMS in the biosciences have driven many of the recent developments in instrumentation, sample handling, and data handling. The increased mass resolving power made available by Orbitrap and mass spectrometry/mass spectrometry (MS/MS) mass analysers gives extremely high chemical specificity and allows the use of gentler ion beam sources that minimise molecular fragmentation as much as possible [60–64].

The flexibility of current SIMS instruments has made it a technique particularly well suited for applications in biomedical science. With the continual improvement in sample preparation and handling (a sample must be compatible with the high-vacuum demands of SIMS), its uses will continue to expand. A selection of some of these applications is presented below.

5.4.1 Early applications and developments

Many of the very early applications of SIMS to biological materials exploited the very good secondary ion yields obtained from the group I and II metals such as Na, K, and Ca, mapping the distribution of their ions (Ca^+, Na^+, and K^+) in bone and between bone and Ti implants [65–69]. Extensive studies showed that during metabolic acidosis, which can induce changes in bone mineral, the organic material was responsible for the buffering of excess H^+ ions and not the bone mineral [70].

Taking advantage of its very high lateral resolving power, the NanoSIMS has been used to great effect in the area of metabolism and drug uptake via isotopically labelled samples, mapping the small molecules of $^{12}C^{14}N$, ^{12}C, and ^{15}N and the elemental species ^{12}C and ^{13}C on a subcellular level. Image ratios of $^{12}C^{14}N/^{12}C\,^{15}N$ were able to show nitrogen fixation in bacteria and protein renewal in kidney tissue, whereas the ^{12}C and/^{13}C image ratios were used to analyse fatty-acid transport in adipocyte lipid droplets [71–73].

In these early examples, elemental secondary ions were collected, and the small molecular ion of CN was typically used as a biological proxy ion. The implementation of the LMIG as the primary ion source in ToF-based SIMS instruments changed the emphasis on the type of SIMS data that could now be collected, going beyond elemental and isotope mapping and determining lipid and metabolite distribution in biological tissue [74–77].

5.4.2 New frontiers

With the variety of SIMS platforms (ToF-SIMS, Orbi-SIMS, and NanoSIMS) that currently exist, their range of application to the biosciences has expanded substantially, allowing researchers to visualise molecular architecture, image biomolecules, and resolve biological processes at a range of length scales [78–81]. With this in mind, only a few applications will be presented here to try to underscore the variety of areas that SIMS is now applied to.

Cancerous tumours have a distinct metabolic process that distinguishes them from healthy tissue. It is a highly dynamic process driven by the tumour cells, as well as transformed stromal infiltrates and immune components. SIMS equipped with an Orbitrap mass analyser was used to capture mass spectrometry images from primary purine tissues with the aim of investigating the pathogenic tissue and differentiating the cell regions from one another [82].

Orbi-SIMS measurements were taken from five different tumour sections. The measurements were able to distinguish between the tumour tissue, the immune cells, and the stromal cells (see figure 5.16) and detected many more peaks from each of

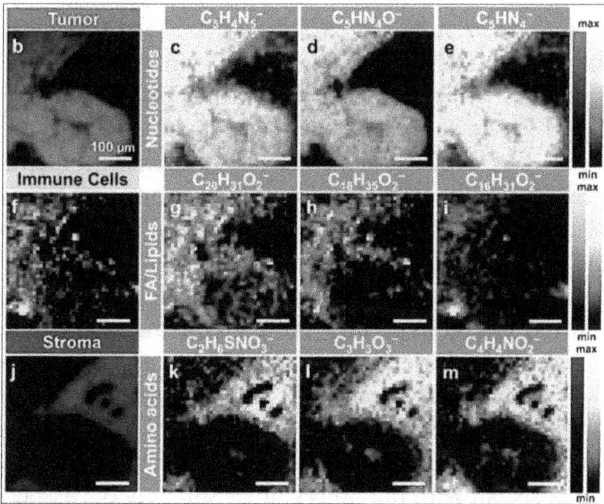

Figure 5.16. (a)Orbi-SIMS images of (b–e) tumour cells characterised by nucleotide fragments, (f–i) immune cells characterised by fatty acids and lipid fragments, and (j–m) stroma cells characterised by amino acids [82] John Wiley & Sons. [© 2024 The Author(s). Chemistry - Methods published by Chemistry Europe and Wiley-VCH GmbH].

the specific tissue types compared to other techniques. In the tumour tissue, 119 distinct mass peaks were found to determine the metabolic profile of the tumour cells. The most intense mass peaks were identified as nucleotide fragments: $C_4N_3^-$, $C_4H_2N_3^-$, $C_4H_2N_4^-$, $C_4H_3N_4^-$, C_5HN^-, $C_5HN_4O^-$, and $C_5H_4N_5^-$. A total of 200 peaks were associated with the immune cells. The immune cells were characterised by fragments of fatty acids and lipids. Finally, the noncancerous stroma cells were found to be distinguished by 72 peaks associated with amino acids.

The Orbi-SIMS analysis was able to clearly differentiate between the three cell types in the tissue samples. Specifically identifying the nucleotide fragments associated with the tumour cells could help in identifying a metabolic pathway for effective cancer therapies.

Another ToF-SIMS study of cancerous tissue closely compared the differences between two analytical techniques, namely principal component analysis (PCA) and multivariate curve resolution (MCR), with the aim of distinguishing between surviving cancer cells after treatment and untreated cells [83]. It was observed that MCR provided a much clearer separation between images of the untreated and treated surviving cells. The major components of the cells, i.e. cytoplasm, nuclei, and lipid droplets, were all resolved. Importantly, the application of MCR to the analytical data revealed the presence of lipid droplets on the surviving cells, indicating that lipids have an important role in the survival of cancer cells.

An interesting application of Orbi-SIMS has been to investigate the biological response to medical device implantation [84]. An important characteristic of a medical device or implant is that it should not cause more problems for the patient and should be benign. In some cases, however, patients may have adverse reactions to a device, such as inflammation, pain, and possibly device failure.

To understand the chemical changes that may occur in the tissue surrounding an implant, two novel polymer coatings that had been identified to elicit different immunological responses were tested: M1 (pCHMA-DMAEMA), a proinflammatory coating, and M2 (pCHMA-iDMA), an anti-inflammatory polymer coating [85]. The coatings were applied to silicone catheters and then placed in mouse models for 28 days. Tissue samples from the area surrounding the catheters were taken and analysed by Orbi-SIMS. The SIMS data clearly identified a chemical difference between the tissues extracted from close to the catheters coated in either M1, the proinflammatory polymer, or M2, the anti-inflammatory polymer.

In the tissue associated with M2, lower mass peaks identified as proteins were observed. This was consistent with a greater production of collagen. Second, histamine and purine peaks were also detected in the tissue samples associated with M2 and not at all in those associated with M1. These compounds are both connected to anti-inflammatory reactions and are clear indicators that the M2 polymer coating influences the local tissue environment. The tissue surrounding the catheter coated in M1 was found to preferentially indicate the presence of glycerol lipids, which was consistent with previous studies showing the association of glycerol lipids with immune cell activity and enhancing inflammation [85].

One area where, perhaps, the application of SIMS has not been so widely used is in the field of environmental science. In the agrochemical field, ToF-SIMS has been

Table 5.2. Pharmaceuticals detected by SIMS on treated biofilms. Reprinted with permission from [89]. Copyright (2023) American Chemical Society.

Compound	Ion peak [*m*/*z*]	Tentative ion species formula
Citalopram	325.1	$C_{20}H_{22}FN_2O^+$
	326.1	C_{19} [13] $CH_{22}FN_2O^+$
Ketoconazole	495.1	$C_{26}H_{28}ClN_4O_4{}^+$
	523.1	$C_{24}H_{29}Cl_2N_4O_5{}^+$
	537.1	$C_{24}H_{27}Cl_2N_4O_6{}^+$
	565.1	$C_{26}H_{31}Cl_2N_4O_6{}^+$
Sertraline	275.0	$C_{16}H_{13}Cl_2{}^+$
	277.0	$C_{16}H_8ClN_3{}^+$
	306.0	$C_{17}H_{18}Cl_2N^+$

applied to imaging biocidal compounds on wheat seeds, and the NanoSIMS has been used to study iron bioavailability in wheat and peas [86–88].

The challenge of maintaining good water quality is made harder by the growing levels of pharmaceuticals within the water system, whether they come from domestic or farming uses. These can cause a number of problems for both ecological and human health. Water treatment plants use biofilms as a way of removing pharmaceuticals from the water system. To identify the localisation of the extracted chemicals and clarify the removal pathways, ToF-SIMS analysis has been carried out on these biofilms [89] (and a list of detected pharmaceutical compounds is highlighted in table 5.2).

References

[1] Dowsett M and Adriaens A 2004 *Nucl. Instr. Meth. Phys. Res.* B **226** 38–52

[2] Keune K, Hoogland F, Boon J J, Peggie D and Higgit C 2009 *Int. J. Mass Spectrom.* **284** 22–34

[3] Bouvier C, Van Nuffel S, Water P and Brunelle A 2021 *J. Mass Spectrom.* **57** e4803

[4] Fearn S, McPhail D S, Morris R J H and Dowsett M G 2006 *Appl. Surf. Sci.* **252** 7070–3

[5] Fricker A L, McPhail D S, Keneghan B and Pretzek B 2017 *npj Herit. Sci.* **5** 28

[6] Ireland T R 2015 Secondary ion mass spectrometry *Encyclopedia of Scientific Dating Methods* ed W J Rink and J W Thompson (Berlin: Springer) 739–40

[7] Vasquez J A and Schmitt A K 2017 *Rev. Mineral. Geochem.* **83** 199–230

[8] Carpenter J D, Stevenson T J, Fraser G W, Bridges J C, Kearsley A T, Chater R J and Hainsworth S V 2007 *J. Geophys. Res. Planets.* **112** E08008

[9] Daividson J, Buseman H, Nittler L R, Alexander C M O' D, Orthous-Daunay F-R, Franchi I A and Hoppe P 2014 *Geochim. Comoschim. Acta* **139** 248–66

[10] Goacher R E, Edwards E A, Yakunin A F, Mimms C A and Master E R 2012 *Anal. Chem.* **84** 4443–51

[11] Charlton D, Costa C, Trindade G F, Hinder S, Watts J F and Bailiey M J 2023 *Sci. Justice* **63** 9–18

[12] Wilson R G, Stevie F A and Magee C W 1989 *Secondary Ion Mass Spectrometry: A Practical Handbook for Depth Profiling and Bulk Impurity Analysis* (New York: Wiley)

[13] Simenas M *et al* 2022 *Phys. Rev. Let.* **129** 117701

[14] Schofield S, Curson N J, Simmons M Y, Ruess F J, Hallam T, Oberbeck L and Clark R G 2003 *Phys. Rev. Lett.* **91** 136104

[15] Stock T J Z *et al* 2020 *ACS Nano* **14** 3316–27

[16] Park J, Yun H J and Hwang H 1998 *Jpn. J. Appl. Phys.* **37** L1347

[17] Cheung K P 2023 *Power Electron. Devices Compon.* **4** 100024

[18] Verwiej J F and Klootwijk J H 1996 *Microelectron. J.* **27** 611–22

[19] Cox H R J, Buckwell M, Ng H W, Mannion D J, Mehonic A, Shearing P R, Fearn S and Kenyon A J 2021 *APL Mater.* **9** 111109

[20] Shard A G, Spencer S J, Smith S A, Havelund R and Gilmore I S 2015 *Int. J. Mass Spectrom.* **377** 599

[21] Cox H R J *et al* 2024 *Adv. Mater.* **36** 2408437

[22] Kloppel K D and Von Bunau J G 1983 *Int. J. Mass Spectrom. Ion Phys.* **49** 11–24

[23] Siddle A, Castle J E, Hultquist G and Tan K 2002 *Surf. Int. Anal.* **33** 807–14

[24] Lutjering G and Williams J C 2003 *Titanium* (Berlin: Springer)

[25] Kim J and Tasan C C 2109 *Int. J. Hydrog. Energy* **44** 6333–43

[26] Kim J, Hall D, Yan H, Shi Y, Jospeh S, Fearn S, Chater R J, Dye D and Tasan C C 2021 *Acta Mater.* **220** 117304

[27] McMahon G, Miller B D and Burke M G 2018 *npj Mater. Degrad.* **2** 2

[28] McMahon G, Miller B D and Burke M G 2020 *Int. J. Hydrog. Energy* **45** 200042–52

[29] Ryan M, Williams D, Chater R J, Hutton B M and MacPhail D S 2002 *Nature* **415** 770–4

[30] Chapman T, Chater R J, Saunders E A, Walker A M, Lindley T C and Dye D 2015 *Corros. Sci.* **96** 87–101

[31] Foss B J, Hardy M C, Child D J, McPhail D S and Shollock B A 2014 *JOM* **66** 2516–24

[32] Mukadam Z *et al* 2023 *Energy Environ. Sci.* **16** 2934–44

[33] Weber M L, Jennings D, Fearn S, Cavallaro A, Prochazka M, Gutsche A *et al* 2024 *Nat. Commun.* **15** 9724

[34] Aranda C A, Alvarez A O, Chivrony V S, Das C, Rai M and Saliba M 2024 *Joule* **8** 241–54

[35] Brugge R H, Chater R J, Kilner J A and Ainara A 2021 *J. Phys.: Energy* **3** 034001

[36] Popovic J 2021 *Front. Nanosci.* **19** 327–59

[37] Brugge R H, Hekselman A K, Cavallaro A, Pesci F M, Chater R J, Kilner J A and Aguadero A 2018 *Chem. Mater.* **30** 3704–13

[38] Zagorac T, Counihan M, Shavandi R, Lee J, Zhang Y, Tepavcevi S and Hanley L 2024 *Micorsc. Microanal.* **30** 200

[39] Shen Z and Fearn S 2024 *J. Electroceram.* **2024** 1–28

[40] Lombardo T, Walther F, Kern C, Moryson Y, Weintraut T, Henss A and Rohnke M 2023 *J. Vac. Sci. Tech.* A **41** 053207

[41] Pesci F M, Brugge R H, Hekselman A K, Cavallaro A, Chater R J and Aguadero A 2018 *J. Mats. Chem.* A **6** 19817–27

[42] Cressa L, Fell J, Pauly C, Hoang Q H, Mucklich F, Hermann H-G, Wirtz T and Eswara S 2022 *Microsc. Microanal.* **28** 1890–5

[43] Walther F, Strauss F, Wu X, Mogwitz B, Hertel J, Sann J, Rohnke M, Brezesinski T and Janek J 2021 *Chem. Mater.* **33** 2110–25

[44] Walther F, Raimund K, Fuchs T, Ohno S, Sann J, Rohnke M, Zeier M and Janek J 2019 *Chem. Mater.* **31** 3745–55
[45] Peled E and Menki S 2017 *J. Electrochem. Soc.* **164** A1703
[46] Westhead O *et al* 2023 *J. Mater. Chem.* A **11** 12746–58
[47] Gaulthier N, Courreges C, Demeaux J, Tesier C and Martinez H 2020 *Appl. Surf. Sci.* **510** 144266
[48] https://gtr.ukri.org/projects?ref=EP%2FP029914%2F1
[49] Chater R J, Smith A J and Cooke G 2016 *J. Vac. Sci. Technol.* B **34** 03H122
[50] Jain V and Pan Y M 2000 *Glass Melt Chemistry and Product Qualification, Centre for Nuclear Waste Regulatory Analyses (CNWRA)* (San Antonio, TX: Nuclear Regulatory Commission) Contract NRC-02-97-009
[51] Lodding A and Van Iseghem P 2001 *J. Nucl. Mater.* **298** 197–202
[52] Chave T, Frugier P, Ayral A and Gin S 2007 *J. Nucl. Mater.* **362** 466–73
[53] Ryan J L 1996 The atmospheric deterioration of glass: studies of decay mechanisms and conservation techniques *PhD Thesis* (University of London)
[54] Fearn S, McPhail D S and Oakley V 2005 *Phys. Chem. Glass.* **46** 505–11
[55] Lodding A, Odelius H, Clark D E and Werme L O 1985 *Mikrochim. Acta Supl.* **11** 145
[56] Wicks G G 1991 Nuclear waste glasses: corrosion behaviour and field tests *Corrosion of Glass, Ceramics, and Superconductors* ed D E Clark and B K Zoitos (Noyes Publication)
[57] Ahmad N E, Fearn S, Jones J R and .Lee W E 2014 *Procedia Mater. Sci.* **7** 230–6
[58] Fearn S 2000 A SIMS based bevel-image technique for the analysis of semiconductor materials *PhD Thesis* (University of London)
[59] Leo B F, Fearn S, Gonzalez-Carter D, Theodoru I, Ruenraroengsak P, Goode A E *et al* 2019 *Anal. Chem.* **91** 11098–107
[60] Passerelli M K *et al* 2017 *Nat. Methods* **14** 1175–83
[61] Hill R, Blenkinsopp P, Thompson S, Vickerman J C and Fletcher J S 2011 *Surf. Int. Anal.* **43** 506–9
[62] Fletcher J S, Rabbani S, Henderson A, Lockyer N P and Vickerman J C 2011 *Rapid Commun. Mass Spec.* **25** 925–32
[63] Bich C, Havelund R, Moellers R, Touboul D, Kollmer F, Niehuis E, Gilmore I S and Brunelle A 2013 *Anal. Chem.* **85** 7745–52
[64] Nilsson K, Karagianni A, Kaya I, Henricsson M and Fletcher J S 2021 *Anal. Bioanal. Chem.* **413** 4181–94
[65] Levi-Setti R, Wang Y L and Crow G 1986 *Appl. Surf. Sci.* **26** 249–64
[66] Bushinsky D A, Chabala J M and Levi-Setti R 1990 *Am. J. Physiol.* **259** E586–92
[67] Lodding A R, Fishcer P M, Odehus H, Noren J G, Sennerby L, Johansson C B, Chabala J M and Levi-Setti R 1990 *Anal. Chim. Acta* **241** 299–314
[68] Chandra S, Fullmer C S, Smith C A, Wasserman R H and Morrison G H 1990 *Proc. Natl Acad. Sci.* **87** 5715–9
[69] Chandra S, Leinhos G M E, Morrison G, Hoch H and H C 2000 *Anal. Chem.* **72** 104A–14A
[70] Bushinsky D A, Gavrilov K L, Chabala J M and Levi-Setti R 2000 *J. Bone Miner. Res.* **15** 2026–32
[71] Kleinfield A M, Kampf J P and Lechene C P 2004 *J. Am. Soc. Mass Spectrom.* **15** 1572–80
[72] Lechene C *et al* 2006 *J. Biol.* **5** 20
[73] Lechene C P, Luyten Y, McMahon G and Distel D 2007 *Science* **317** 1563–6

[74] Debois D, Bralet M-P, LeNaour F, Brunelle A and Laprevote O 2009 *Anal. Chem.* **81** 2823–31
[75] Denbigh J J and Locker N P 2015 *Mater. Sci. Technol.* **31** 137–47
[76] Lazar A N *et al* 2013 *Acta Neuropathol.* **125** 133–44
[77] Kezutyte T, Desbenoit N, Brunelle A and Briedis V 2013 *Biointerphases* **8** 1–8
[78] Watrous J D and Doreestein P C 2011 *Nat. Rev.* **9** 683–94
[79] Agui-Gonzalez P, Jahen S and Phan N T N 2019 *J. Anal. At. Spectrom.* **34** 1355
[80] Massonet P and Heeren R M 2019 *J. Anal. At. Spectrom.* **34** 2217
[81] Jia F, Zhao X and Zhao Y 2023 *Front.Chem.* **11** 1237408
[82] Kern C, Scherer A, Gambs L, Yuneva M, Walczak H, Liccardi G *et al* 2024 *Chem. Methods* **4** e202400008
[83] Manaprasertsak A, Rydeberg R, Wu Q, Slyusarenko M, Carroll C, Amend S R, Mohlin S, Pienta K J, Malmberg P and Hammarlumnd E U 2025 *Anal. Methods* **17** 2263–72
[84] Suvannapruk W, Fisher L E, Luckett J C, Edney M K, Kotowska A M, Kim D-H, Scurr D J, Ghaemmaghami A M and Alexander M R 2024 *Adv. Sci.* **11** 2306000
[85] Rostam H, Fisher L E, Hook A, Burroughs L, Luckett J C, Figueredo G P, Mbadugha C, Teo A C K *et al* 2020 *Matter* **2** 1564–81
[86] Converso V, Fearn S, Ware E, McPhail D S, Flemming A J and Bundy J G 2017 *Sci. Rep.* **7** 10728
[87] Moore K L, Zhao F-J, Gritsch C S, Tasi P, Hawkesford M J, McGrath S P, Shewry P R and Grovenor C R M 2012 *J. Ceral Sci.* **55** 183–7
[88] Moore K L, Rodriguez-Ramiro I, Jones E R, Rodriguez-Celma J, Halsey K *et al* 2018 *Sci. Rep.* **8** 6865
[89] Burzio C, Mohammadi A S, Malmberg P, Modin O, Persoson F and Wilen B-M 2023 *Environ. Sci. Technol.* **57** 7431–41

Chapter 6

Outlook

Secondary ion mass spectrometry is an extremely versatile analytical tool that can be used for the characterisation of a wide range of materials. Its unique combination of high-resolution depth profiling, highly resolved ion imaging, and mass spectrometry with very high sensitivity can generate a huge amount of information with regard to elemental, molecular, and chemical composition, not only in two dimensions but also in three.

As previously observed, much instrument development has arisen through the analytical needs of the materials being studied. This is probably most evident in the area of biological measurements. The need for enhanced sensitivity and greater mass resolving power has led to the development of Orbitrap SIMS and new ion beam sources such as gas cluster ion beams (GCIBs) that allow for the desorption of molecular ions. Further development of novel ion sources will continue to push the boundaries of analyses in biological measurements. Currently, the sensitivity and spatial resolving capabilities of GCIBs are limited, as they are not high-brightness sources and cannot be focussed to the small spot sizes needed for submicron imaging. The development of reactive GCIBs would improve ion yield and help enhance detection levels. Continued improvement in mass peak assignment would also be beneficial to advance precise metabolite and biomolecule identification.

Plasma focussed ion beams (PFIBs) are now being applied to instrumentation. These monoatomic and diatomic sources are high-brightness sources that can achieve lateral resolving powers of $\sim$50 nm. The application of these sources to the analysis of quantum devices and battery materials will be hugely beneficial. In particular, advances in understanding and optimising Li and N battery operation will come from detailed 3D nanoscale imaging made possible by these sources. PFIBs possess high milling rates which can be exploited to perform *in situ* sample preparation, revealing buried features and interfaces under vacuum, which is ideal for air-sensitive samples.

doi:10.1088/978-0-7503-3331-3ch6

Future advances will be made in data handling, particularly with regard to the large and complex data sets obtained from biological samples. The application and development of machine learning and AI will be helpful resources in overcoming issues associated with matrix effects, improving peak assignments, and recognising patterns in images.

Finally, correlated imaging will also develop, as cryo-sample preparation common to both SIMS and high-resolution microscopy, coupled with cryo-transfer, will facilitate obtaining chemical and morphological information from the same sample.

Secondary ion mass spectrometry will continue to be a fundamental tool in advanced materials research. Any future advances will only further enhance its already wide-ranging unique capabilities.

www.ingramcontent.com/pod-product-compliance
Lightning Source LLC
Chambersburg PA
CBHW082105210326
41599CB00033B/6595

* 9 7 8 0 7 5 0 3 3 3 3 2 0 *